The Next Four Years of Trump

Artificial Intelligence Predicts the 2025-2029 Presidency

Not for Nothing Books

The Next Four Years of Trump

ISBN: ISBN: 978-1-958157-32-9 (Print)
ISBN: 978-1-958157-33-6 (Ebook)

Any references to historical events, real people, or real places are factually verifiable whereas the analytical predictions based on said events, people and places were created with Artificial Intelligence and are not the work or opinion of the publisher or anyone involved in the creation or publication of this work.

Front cover image by Not for Nothing Books.

Printed by Not for Nothing Books, in the United States of America.

First printing edition 2025.

Table of Contents

Disclaimer about Artificial Intelligence

Foreword by Thorton Maldonado

PART I - Donald J. Trump's Early Life, Family Struggles and Controversial Deals

PART II – Trump's Playbook: Project 2025

PART III – Project 2025's Components, Principles, Goals & Agenda

PART IV – How Trump Would Take Reins of the Administrative State

PART V – Defunding and Cutting Major Institutions like Education

PART VI – Project 2025's Plan for Your Wallet and the Economy

PART VII – Placing Independent Agencies Under Direct Control of Trump

PART VIII - Changes to the Executive Office of the President

PART IX - Other Democratic Guardrails in Peril

PART X - Justice System and Law Enforcement

PART XI – Impact on Health and Human Services

PART XII – More Healthcare Reforms (Medicaid, Medicare, Etc.)

PART XIII – Labor, Unemployment, Kids Working Dangerous Jobs & Other Labor Reforms

PART XIV – Education and Childcare

PART XV – Civil Rights / Gender Identity/ Sexual Orientation

PART XVI - Agriculture and Nutrition

PART XVII - Homeland Security

PART XVIII – Military, Defense and Foreign Adversaries

PART XIX – International Relations & Tariff Warfare

PART XX - Environmental Deregulation and Defunding

PART XXI – Commerce, Treasury and Tax Reforms

PART XXII – Immigration

PART XXIII - Transportation and Infrastructure

PART XXIV - Elections

PART XXV - Culture War

PART XXVI – Bringing Down Inflation

PART XXVII – Trump's First 100 Days (2025)

PART XXVIII – Trump's First Term (2017-2021)

PART XXIX – The Next Four Years under Trump

PART XXX – Trump's Malignant Narcissism, Sociopathy and Cognitive Decline

PART XXXI – Trump's 2016 Election and Russia's Interference

PART XXXII - Chapter: Trump's Impeachment Trials

PART XXXIII – Trump's 34 Felony Convictions

PART XXXIV – Trump's Controversial and Criminal Acts and Ideas

PART XXXV – Trump's Relationship with Sex Slave Trafficker Jeffrey Epstein

PART XXXVI – Trump's Relationship with Nazi-saluting billionaire oligarch Elon Musk

PART XXXVII – Trump's Denial of 2016 Election Loss and the Capitol Riot

PART XXXVIII – A Cautionary Note (The Rise of Fascism in Nazi Germany)

PART XXXIX – Best and Worst Case Scenarios for Trump's Second Presidential Term

PART XL – The Last Word: Donald J. Trump Quotes

Disclaimer about Artificial Intelligence

No ideas or opinions herein regarding the publication's thesis and findings are that of the publisher, writer or distributors or any parties involved in its creation or dissemination. All predictions were produced through prompt engineering with Artificial Intelligence, or A.I. (i.e. Generative Pre-Trained Transformers or GPTs and Large Language Models or LLMs) with RAG (Retrieval-Augmented Generation) that combines LLMs with information retrieval systems. The goal of RAG is to create more accurate and relevant text. Prompt engineering involved direct prompts and queries to the LLM through the GPT regarding source material that applied to the central question of this book: the presidential term over next four years (2025-2029). In this way, the sources and prompts were impartial and used only the words and actions of those referenced by Donald J. Trump himself or those involved in his administration (such as those involved in the creation of Project 2025 like Russell Vought). There were no alternate facts, media or political bias, woke or DEI agenda, etc. being the analysis and predictions were done via Artificial Intelligence.

For example, the Artificial Intelligence predicted with these prompts:

Prompt 1:

Based on Trump saying he will be a dictator day 1 of his second term in presidential office and him using Project 2025 as a playbook, predict the next four years of his administration for 2025-2029.

Artificial Intelligence:

Project 2025 proposes that the entire federal bureaucracy, including independent agencies such as the Department of Justice, be placed under direct presidential control - a controversial idea known as "unitary executive theory". In practice, that would streamline decision-making, allowing the president to directly implement policies in a number of areas. The proposals also call for eliminating job protections for thousands of government employees, who could then be replaced by political appointees. The document labels the FBI a "bloated, arrogant, increasingly lawless organization". It calls for drastic overhauls of the agency and several others, as well as the complete elimination of the Department of Education.

Shortly after being sworn in, Trump moved to eliminate job protections for career civil servants, and freeze federal spending. Through Elon Musk and the Department of Government Efficiency (which is not an actual government agency), the White House has moved to chop billions in federal spending, although the details and legal status of the cuts are hazy at best. DOGE is not an official government department, but rather an outside team advising Trump with broad authority from the president. It's clear however that Trump intends to take a sledgehammer to the federal government as it currently stands - a goal broadly in line with Project 2025 suggestions. Abortion and family The mentions

of abortion in Project 2025 - there are about 200 of them - have sparked some of the most contentious debate. The document does not call for an outright nationwide abortion ban, and Trump says he would not sign such a law. However, it proposes withdrawing the abortion pill mifepristone from the market, and using existing but little-enforced laws to stop the drug being sent through the post.

The document proposes new data collection efforts on abortion and more generally suggests that the department of Health and Human Services should "maintain a biblically based, social science-reinforced definition of marriage and family". Trump, by contrast, has generally said that abortion laws should mostly be left to individual states. However, during confirmation hearings, Trump's nominee for health secretary, Robert F Kennedy Jr, said the president had ordered him to examine the safety record of mifepristone and left open the possibility of further regulation of the drug. Trump also issued an executive order designed to stop federal funds being used for abortion, a move that was outlined in detail in the Project 2025 document. Immigration EPA Migrants at the US southern border wall in Juarez City, Mexico EPA Increased funding for a wall on the US-Mexico border - one of Trump's signature proposals in 2016 - is proposed in the document. But Trump's signature immigration policy - a pledge to deport millions of undocumented immigrants - is not spelled out in any detail in Project 2025. The document does include language calling on Trump to "thoroughly enforce immigration laws". But in the main chapter dealing with immigration, Project 2025 authors suggest dismantling the Department of Homeland Security and combining it with other immigration enforcement units in other agencies, creating a much larger and more powerful border policing operation.

Other proposals include eliminating visa categories for crime and human trafficking victims, increasing fees on immigrants and allowing fast-tracked applications for migrants who pay a premium. But it was mass deportations - not a bureaucratic shuffle, visa changes or a longer, taller border wall - that was Trump's top pitch to voters. On this issue, his administration promises to go in a slightly different direction - and potentially much further - than the Project 2025 proposals. Anthony Zurcher: Result hands Trump free rein Energy, climate and trade Energy policy is a broad area of agreement between Trump and the Project 2025 proposals, summed up by one of the president's campaign slogans: "Drill, baby, drill". The new administration wants to ramp up fossil fuel production and has taken the US out of the Paris Agreement on climate change, which seeks to limit emissions and global warming. Project 2025 proposes slashing federal money for research and investment in renewable energy, and calls for the next president to "stop the war on oil and natural gas" - ideas that the Trump campaign has enthusiastically taken up. The document sets out two competing visions on tariffs: one suggesting boosting free trade and another pro-tariff position. Trump has clearly sided with the latter camp, announcing import taxes targeting Canada, Mexico and China. The economic advisers of Project 2025 suggest that a second Trump administration should slash corporate and income taxes, abolish the Federal Reserve and even consider a return to gold-backed currency. While the president has made comments about proposals in some of these areas, the economic talk in the early days of his administration has been dominated by tariffs.

Education, tech and DEI Almost immediately upon taking office, Trump moved to end diversity, equity and inclusion programs and decreed that government departments would recognize only two genders. Those moves are broadly in line with Project 2025, which took aim at DEI and gender

terminology as part of what it describes as a wider crackdown on "woke" ideology. The document also calls for greater school choice - essentially subsidizing religious and private schools with public funds - which was also the subject of an early Trump executive order. And it calls for abolishing the Department of Education, another idea that Trump has signaled he supports. In other proposals, Project 2025 suggests banning pornography and shutting down tech and telecoms companies that allow access to adult material. This has so far not been a focus of the new administration, which has drawn support from a number of top tech bosses. Trump's views on the tech industry have regularly shifted, and don't appear to have much to do with sexual content.

Prompt 2:

What would each year look like based on Prompt 1?

Artificial Intelligence:

If Donald Trump were to win a second term in office and fully embrace the proposals outlined in *Project 2025*, we would likely see an extremely transformative and controversial period in American history. The project itself is a sweeping plan to reshape the federal government, society, and the economy along more conservative, right-wing lines. Here's a prediction for what the next four years of Trump's administration might look like, based on the project's principles and proposals.

Year 1: Reorganization and Overhaul of the Federal Government

Trump would likely start his second term by implementing drastic changes to the structure of the federal bureaucracy, many of which are in line with the "unitary executive theory" outlined in *Project 2025*. His administration would seek to centralize power in the executive branch, diminishing the independence of agencies like the Department of Justice (DOJ) and the Federal Bureau of Investigation (FBI). In practice, this would mean shifting control over all federal agencies directly to the White House, allowing Trump to directly influence policy decisions without significant checks from other branches of government.

- **Firing career civil servants: He would likely begin by eliminating the job protections of career federal employees, replacing many with political appointees loyal to his administration. This would be a key part of his strategy to push his agenda more swiftly and efficiently.**
- **Freezing federal spending: Trump's administration would also freeze or slash federal spending, possibly to align with *Project 2025's* goals of cutting billions of dollars in government waste. The Department of Government Efficiency (DOGE), an outside team with broad presidential authority, would play a major role in this.**

Year 2: Social and Cultural Battles Intensify

Trump would likely take a more direct approach on issues related to abortion, family values, and education, which are significant themes in *Project 2025*.

- **Abortion and Family: Trump would continue his push to restrict abortion access, including potentially moving to limit access to mifepristone, the abortion pill. Although *Project***

2025 does not call for a full nationwide ban, Trump may use executive orders to restrict abortion in specific ways, including limiting federal funding for abortion services. In tandem with this, Trump's administration would likely expand efforts to enforce a more conservative definition of marriage and family, drawing heavily on Christian social values.

- Education: Trump would likely prioritize school choice policies, redirecting public funds to private and religious schools, and possibly moving to abolish the Department of Education, a policy long supported by conservative groups. He would likely push for the "de-wokeifying" of education, using *Project 2025* to target diversity, equity, and inclusion (DEI) programs.
- "Woke" Agenda: Alongside this, Trump would implement crackdowns on what he deems "woke" culture, enforcing laws that restrict gender recognition to two categories, likely creating a cultural flashpoint. This might include revising federal policies related to gender, especially in the military and healthcare.

Year 3: Immigration Overhaul and Border Security

Immigration would likely remain one of Trump's primary focus areas, with *Project 2025* suggesting a reorganization of the Department of Homeland Security (DHS) and increased efforts at border enforcement. However, Trump's policy would likely extend beyond bureaucratic reorganizations.

- Mass Deportations: While *Project 2025* calls for enforcement of existing immigration laws, Trump's campaign promises have revolved around mass deportations. This could be a key policy during his second term, likely resulting in a significant increase in deportations of undocumented immigrants, especially from communities with high numbers of undocumented workers.

- Immigration Visa Changes: The Trump administration would continue to push for more stringent immigration laws, eliminating visa categories for crime victims or individuals involved in human trafficking. These moves would likely increase barriers for migrants seeking asylum and force many to enter through more dangerous routes.

Year 4: Economic and Environmental Policies in Focus

Trump's economic policies would likely be an extension of his previous administration's agenda, with an emphasis on deregulation, fossil fuel expansion, and tariffs.

- Energy and Environment: Trump's administration would almost certainly ramp up fossil fuel production, continuing the rollback of environmental regulations and pushing for more drilling, particularly in federal lands. A full withdrawal from global climate agreements, such as the Paris Accord, would likely be one of his first moves.
- Trade Policies: Trump's position on trade would likely remain protectionist, with the continuation of tariffs on key trading partners such as China, Mexico, and Canada. He may also push for a more isolationist trade agenda, possibly even revisiting ideas like tariffs on European Union goods.
- Corporate Tax Cuts and the Federal Reserve: There may be efforts to cut corporate taxes again, alongside moves to reduce income taxes. Trump might also pursue more drastic economic changes, like dismantling the Federal Reserve and exploring alternatives like a gold-backed currency—though such a move would face significant opposition.

Overall Tone: A Centralized, Polarizing Administration

Trump's second term, under the influence of *Project 2025*, would likely see a highly centralized executive branch with immense power concentrated in the hands of the president. The rhetoric and policies would be deeply divisive, with harsh opposition likely from progressives, many moderate Republicans, and certain sections of the business community. Public opinion would likely become increasingly polarized as these changes unfold.

There would be a concerted push to "reclaim" American identity in the cultural and political spheres, with significant pushback from the left over issues like abortion access, LGBTQ+ rights, and racial equality policies. Meanwhile, the economy would likely experience a period of turmoil, especially as Trump's more radical economic proposals—such as dismantling the Federal Reserve—are explored.

The next four years could see a strong clash between Trump's vision for America and those advocating for the traditional structure of government and civil rights protections, making his second term one of the most contentious in modern American history.

Being Trump is trying to enact all the Project 2025 proposals on his first few weeks in office, the prompt changed to this with the following results:

Prompt 3:

Redo with Trump trying to do all of the Project 2025 proposals in the first few days of office; still predict what happen in the next four years.

Artificial Intelligence:

Based on your original question about Trump attempting to implement all of the *Project 2025* proposals in the first few days of his second term, I can still provide a detailed prediction of what the next four years might look like.

Year 1: Immediate Reorganization and Political Upheaval

1. Unitary Executive Theory and Bureaucratic Overhaul: Trump would start his second term with a sweeping consolidation of power. The first few days of his administration would likely see immediate executive orders to implement the "unitary executive theory," aiming to centralize control over all federal agencies. His first significant moves would likely involve stripping independence from key agencies such as the Department of Justice (DOJ) and the FBI, which would be put directly under presidential control. This would make it easier for Trump to bypass traditional checks and balances, allowing him to implement policies without much external oversight.

- Mass firings and reshuffling: Trump would likely eliminate job protections for civil servants across the federal government, firing career employees and replacing them with loyal political appointees. This would be a radical change and could face significant backlash from public service unions and liberal groups.

2. Economic and Fiscal Moves: In the first few days, Trump would take action to freeze federal spending and begin drastic cuts to government programs. His team, likely working alongside advisers from Project 2025 like those involved in the fake Department of Government Efficiency (DOGE), would push for immediate budget freezes and cuts to social safety net programs. This could trigger a government shutdown if Congress doesn't comply with his demands, leading to a major political crisis.

- **Massive deregulation: Trump would immediately start dismantling federal regulations, particularly those targeting fossil fuel industries and environmental protections. This would include rolling back Obama-era climate initiatives and prioritizing oil and gas drilling.**

Year 2: Social and Cultural Transformation

3. Cultural Battles: Trump would likely prioritize cultural and social issues early in his second term. In the first few days, he would likely issue executive orders to abolish diversity, equity, and inclusion (DEI) programs within the federal government, targeting what *Project 2025* describes as "woke" culture. This would mean actions like banning federal recognition of non-binary genders, potentially reshaping federal programs and healthcare guidelines.

- **Gender and Marriage Definitions: Trump would likely impose strict policies regarding gender definitions, restricting official recognition to only male and female categories. He might also push for policies that enforce a more "biblically based" definition of marriage, leading to potential conflicts with** states and businesses that support more progressive positions.

4. Education and Family Policy: Trump's administration would likely push for the abolition of the Department of Education, moving swiftly to direct federal funds toward school choice programs, including private and religious schools. The shift would likely be rapid, and Trump could use his executive authority to incentivize states to adopt similar policies.

- Defunding abortion services: Trump would probably issue an executive order aimed at defunding abortion services across the federal system. This would not result in a nationwide ban on abortion but would restrict access to abortion services federally, with a specific focus on limiting access to abortion pills like mifepristone.

Year 3: Immigration and Border Security

5. Immigration Overhaul: Trump's administration would likely enact sweeping changes to the country's immigration laws and enforcement systems in line with *Project 2025*. The first 100 days would likely see the implementation of executive orders to:

- Reorganize DHS and Border Control: Trump would likely dismantle the Department of Homeland Security (DHS), merging it with other immigration enforcement agencies, creating a more powerful agency with broad authority over immigration policy and enforcement.
- Mass deportations: Trump would likely order the mass deportation of undocumented immigrants, which would

spark massive legal and humanitarian crises. Large numbers of migrant families could be detained and removed from the U.S., causing public outrage and mass protests.

- Expanded Border Wall and Increased Funding: Trump would likely push for increased funding to expand the southern border wall, a key promise from his first term, possibly facing opposition from Congress but trying to bypass it through emergency declarations.

Year 4: Economic Isolationism and Global Realignment

6. Trade Wars and Tariffs: Trump would likely escalate his protectionist trade policies, using tariffs as leverage against countries like China, Mexico, and Canada. His administration would aggressively pursue tariffs on imports to boost domestic production, likely causing trade tensions with allied countries and provoking retaliation.

- Corporate Tax Cuts and Economic Reforms: Trump would push for sweeping corporate tax cuts, aiming to stimulate the economy with supply-side policies. He would also likely explore proposals to abolish the Federal Reserve and implement alternative currency systems, potentially including a gold-backed currency. This would cause significant instability in the global financial system and face strong opposition from financial institutions and moderate Republicans.

7. Energy Policy: Trump's administration would prioritize the expansion of fossil fuel industries, ramping up oil drilling and natural gas extraction. *Project 2025* emphasizes ending the "war on oil," so Trump would likely move to reverse Obama-era environmental protections, possibly pulling out of

additional international climate agreements, while continuing to downplay the effects of climate change.

8. Tech Crackdown and Regulation: Despite the growing role of technology in American society, Trump would push forward with policies designed to restrict tech companies, possibly through regulations targeting content moderation, or attempting to break up large companies under the guise of curbing "big tech." The administration would likely continue its assault on "woke" tech, banning certain content (such as pornography) and requiring tech platforms to remove "harmful" content. This could spark major pushback from tech companies and their users.

Political and Legal Fallout

By the end of the four-year term, Trump's second administration would likely face significant opposition from various sectors of American society, including progressive activists, labor unions, civil rights organizations, and some establishment Republicans. Legal challenges to many of his actions, especially his restructuring of the executive branch and his moves to limit abortion access, could clog the courts. His approval ratings would likely be highly polarized, with staunch supporters praising his actions as fulfilling campaign promises, while his critics decrying them as authoritarian.

In summary, Trump's second term, assuming he moves swiftly to implement *Project 2025*, would be characterized by an aggressive consolidation of power, sweeping changes in social, economic, and cultural policies, and significant conflict with both domestic and international actors. The administration would be highly controversial, setting the stage for a deeply divided America in the years to come.

Let's unpack all of this including what Project 2025 is, what Trump may do in the worst case scenarios and how Artificial Intelligence analyzed and predicted the next four years under this presidential administration. First, a forward then a brief biography of Trump.

Foreword

Dear Donald J. Trump,

My name is Thorton J. Maldonado. I am a U.S. citizen born in New York. My family was born in America too going back many generations, probably as many as yours if not more. We are a military family too. My great-grandfather fought in World War I against tyranny, and my grandfather fought against the Nazis in World War II. Both were Marines. My father was in the Air Force. I did not have the honor to serve, but it was not about bone spurs like you used to dodge the Vietnam Draft. I have served my country and its communities instead as a teacher for about 25 years now in some very deep and dangerous trenches at times.

I grew up in the 1980s in New York City. I believe the first time I became aware of you was on a McDonald's commercial. Then it was the Daily News and NY Post front covers about you and your wives and divorces and your messy real estate and casino dealings. Rappers would reference you as some kind of gold standard in their songs to make it seem like they were rich too. I even read your book

The Art of the Deal fully aware that it was ghost written (or just written and you put your name on it like you did with many buildings in the city). I even watched some of The Apprentice. You were part of pop culture, so I rolled my eyes at the joke it all was like when you were wrestling Vince McMahon for some WWE pay-per-view. I always knew you were a con man, but your bankruptcies and scandals were boring and nonconsequential beyond whomever your defrauded or hurt. Little did I know then what nightmares may come.

I was not surprised you ran for office. Obama made a joke about you after you racially pestered him about his birth certificate. I am surprised that the media and your MAGA cultists sane-washed and normalized you so much as you went around whining like a little brat while insulting everyone (it got you more coverage to your credit). But I always knew that there would be backlash once Barack Obama became president. Too much racism in the United States for there not to be a reactionary tide of white privilege and misdirected, socioeconomic grievances being stockpiled for decades. I was surprised that someone with no experience and no platform other than a wrestling gimmick slogan like "Make America Great Again" would get everyone to not focus on your grift and the obvious corruption with Russian interference. But voters are like that (those who bother to vote at all). What was it said of the people only wanting bread and circus?

I was not surprised you tried to stage an insurrection and put your own MAGA cultists in harm's way. They could have all been hurt or worse on January 6, 2021 (not to mention the cops who were). You have been found guilty of many crimes, and so you only know how to resort to crime. And criminals surround you, which means you will never have true loyalty. Underneath it all, you know you are forever unloved. But

when you turn on an entire nation over that malignant narcissism gone awry after being legitimately rejected, that is treason. You never faced consequences to that or any of your rape-y behavior, and I knew you would be reelected eventually though you were defeated thoroughly by Biden. I still wonder where all those votes went for Harris. As you have said, the elections are rigged.

Anyway, we find ourselves in the present moment now in 2025. I began to use my expertise and skills in Artificial Intelligence, Machine Learning and Data Science to figure out what your "concepts of a plan" were going be for your second term in the presidential office. I prompt engineered available LLMs and GPTs while also building a RAG to get updated information. Ask Sam Altman of OpenAI to explain it to you before he takes all that taxpayer money for an AI Stargate project that even your puppet master Elon Musk said was a fraudulent dead stick in the water.

The analysis and prediction in this book basically followed the Project 2025 blueprints (that you denied knowing about) down a rabbit hole. Now virtually every executive order you sign is torn from that blueprint. I personally think that Elon Musk will use then lose you once he gets all the information and access you are letting him have to payment systems like Social Security and Medicare (not to mention nuclear codes). Even his toddler son said you were "not the president," "need to go" and you should "shush your mouth" in the Oval Office while Musk gave a propaganda disinformation speech instead of you. I actually felt embarrassed for you as I was horrified by the shadow king rendering you small and ineffectual at the Resolute Desk despite how much you bully everyone else.

You will make yourself obsolete just as you are trying to do with Congress by bullying them out of their constitutional power to make laws and control financing budgets. You did the same thing to your MAGA voters that you are selling out to the oligarchs and your own mad whims (not to mention a cryptocurrency scam). You did warn everyone that you would be a dictator on day one. Many claim to not take you literal.

I never believed anyone took you figuratively when you said things literally, especially when it was about hurting people or overthrowing the government. Your MAGA death cultists are not that smart, and they would rather remain ignorant so long as you deliver on their crackpot agendas of white supremacy and Christian nationalism and "draining the swamp." They are not going to be distracted for much longer as you kill their government benefits, the economy as a whole and our place as a world leader. It is still some warped white privilege that they are poisoned with even though you and the rest of the rich, kleptocratic sociopaths and sycophants (effing with the U.S. government and American Dream for everyone) think that the MAGA cultists are just white trash (Musk said as much and swore to wage war on them). You said you love the poorly uneducated, and they cheered since they have been beaten down for so many years that they only feel loved when abused. They deserve so much better than you, but freedom means people can choose to destroy democracy and live in slavery.

You won, but it is pyrrhic in nature. You are a success, but only in that you gamed the people and got away with so much crime and corruption (not to mention a coup). You will also fail. No one is loyal to you. But you are not to them either. You already sold them out to the highest bidder or the dirtiest bastard. They know your annexing of other countries like Gaza is going to lead to war and terrorism and destruction. But that kind of seems to be the theme of your

second term. Tariffs alone will kill jobs, raise grocery prices and make the working class and poor subject to suffering and slavery. It will destabilize the stock market and cause so many to lose their investments and savings (except the oligarchs shorting the blatant market manipulations and engineered black swan events). The problem is when you win as the president by doing these unconstitutional and godless acts, the nation itself is the loser. All those people losing their jobs are evidence; all those families broken apart are proof; all our allies turning on us and all our enemies sharpening their teeth are just a furtherance.

I really do wish you would use your platform to help everyone, but you already said you will not bring the country together and help liberals, Democrats, progressives, BIPOC people, LGBTQIA+ people, disabled and special needs people or even children (even those who really on the Department of Education for learning and food). Not even your own MAGA cultists. All you will do is try to get power and get revenge on anyone who you think was against you. You have the opportunity to get the best and brightest to help you be a great president for the entire country, but you get manipulated by praise, money and your demagogy. I saw at Margo Largo that you got a gift of a goat statue made out of fake money with your photo on the bills with a caption reading "In Trump We Trust." You are getting so played. Even if you get all that power, one of your "loyalists" will steal it from you or sell it to one of the enemies of America or just let slip chaos and carnage. You really do not even know what your administration members are up to and who they really work for or how they will go to the highest bidder like Musk or Putin. That is what happens when you make criminals your inner circle, patent corruption as your legacy and pardon insurrectionists who want to overthrow all governments eventually.

We all deserve so much better than that. I just keep seeing you all punked out at the Resolute Desk with a rich bitch oligarch's son wiping snot on your throne and all over our proud traditions as Americans. Just remember this little historical factoid: There has never been co-dictators. And there never will be. So you may be "dictator on day 1," but that does not mean you will be one by day 100 or day 1000 or . . . Shush your mouth?

I waited for a few weeks after your inauguration in 2025 to publish this book to give you the benefit of the doubt, but you are doing what the Project 2025 playbook has laid out and what Artificial Intelligence has predicted herein. You are also pursuing imperialist ideations and fascist/ racist policies that even your own staff know are insane and your Fox News propaganda ministry cannot pretzel-logic into some alternate fact or "genius" negotiation tactic. You already have said you want to own Gaza and ethnic cleanse Palestinians, or that you were abandoning your campaign promise to bring down the price of groceries and not worsen inflation (which you have already with your tariffs or the threat of them). I could have named this book *Dictator on Day One*, but I followed A.I., which predicts you will wage war on the American people and our institutions and the very U.S. Constitution that I love and honor. I even could have taken out the "U" on the front cover of this book and put in a swastika. I remained objective though the predictions chilled me to the bone. I opted to not be salacious, but rather, to go along with the predictions of the GPTs, LLMs and the other machine learning, data science and artificial intelligence tools and techniques implemented in trying to decipher what you yourself and your "loyalists" cannot probably fathom or foresee. Lest we forget, Vice President JD Vance compared you once to "American Hitler" while your administration's secretaries Mark Rubio and Tulsi Gabbard declared you were unfit for the presidency and Musk even said in 2022 that you should "hang up your hat and sail into the sunset." That was around the time you called him "a liar."

Maybe you and all those who "love" you should read this book too, so you know what the eff you are actually doing.

Then again you never cared and never will. Hopefully, you, at least, wiped Musk's son's boogers off the Resolute Desk where wars were won, millions, if not billions, of lives were saved and freed and the greatest nation in history rose. I do not need A.I. to predict what you really did though nor what you will thus do next.

All hail the chief,
Thorton J. Maldonado

PART I - Donald J. Trump's Early Life, Family Struggles and Controversial Deals

Donald Trump was born on June 14, 1946, in Queens, New York, to Fred Trump, a wealthy real estate developer, and Mary MacLeod Trump. While the Trumps were financially successful, Donald's relationship with his father was complex and often strained, which would leave a lasting impact on him throughout his life and career.

1. A Tough Childhood: Strained Relationship with Fred Trump

Donald Trump's father, Fred, was a hard-nosed businessman, and he didn't hesitate to impose his will on his children. Fred was known to be demanding, often focusing more on business than on nurturing his children's emotional well-being. From a young age, Donald was expected to take over the family business, and Fred's tough love approach created an atmosphere of high expectations and little affection. Fred's controlling nature and the constant pressure to succeed led to a difficult dynamic between the two. According to Donald himself, his father's expectations were immense, and he believed Fred was never truly proud of his accomplishments, despite his financial success.

Fred Trump's methods of controlling Donald's ambitions, including limiting his autonomy, reportedly shaped the younger Trump's desire to break free and prove himself. Trump has often recounted how his father made him work hard, but also pushed him to do things he didn't want to do, which built resentment. His relationship with his siblings was also influenced by their father's harshness and competition within the family.

The Influence of Fred Trump on Donald's Path to Power

Fred Trump, a figure driven by ambition, discipline, and an unwavering desire to win, set the stage for his son Donald's worldview and eventual rise to political prominence. A real estate developer known for his political savvy, Fred was notorious for using connections, clever schemes, and an endless work ethic to manipulate the system for personal gain. While he rarely took time to offer emotional guidance or parenting advice, Fred expected Donald to absorb life lessons "by osmosis," simply by observing him at work. But the lessons Donald learned were far from conventional—they were rooted in manipulation, ambition, and self-interest.

One pivotal moment in Donald's early life occurred when he was seven years old. His father faced a U.S. Senate investigation into housing abuses involving government programs intended for veterans and middle-class families. The investigation uncovered how Fred Trump had taken advantage of government loans and rent subsidies by inflating construction costs and collecting rent that wasn't owed. His actions, though legal, were a clear exploitation of the system—an approach to business that would come to define Donald's own dealings later in life. Fred's unapologetic attitude toward these transgressions was formative for Donald. Even as federal officials were outraged and President Eisenhower decried Fred's actions, Donald absorbed the idea that breaking the rules could lead to great financial rewards.

Fred Trump's influence extended beyond business into Donald's moral and ethical framework. His father was strict and demanding, yet emotionally distant, expecting his children to meet his high standards without the kind of guidance or affection that typically accompanies parental relationships. Donald often spoke of his father as tough and unyielding—a man who was never satisfied and never let go of his expectations. These traits would shape Donald's own character, pushing him toward a life where the relentless pursuit of success became his primary focus, leaving little room for compassion or humility.

Fred's teachings about the value of power and success in life were echoed in the harsh, militaristic environment of the New York Military Academy (NYMA), where Donald was sent at age twelve after he had been deemed unmanageable by his father. At NYMA, a school defined by competition, physical toughness, and a strict hierarchy, Donald encountered an environment that reflected the values his father instilled in him. The military school was notorious for its culture of physical abuse and bullying, where cadets were expected to learn survival skills through violence and dominance. It was here that Donald Trump first learned the brutal code of survival-of-the-fittest and developed an appreciation for the kind of strength that comes with dominance over others.

Among the influential figures at NYMA, Sergeant Theodore Dobias became a key mentor to Donald. A World War II combat veteran, Dobias was known for his tough and sometimes cruel methods of discipline. Under Dobias's tutelage, Trump was exposed to a leadership style that emphasized dominance, strength, and unquestioning obedience. He absorbed the mantra, "winning isn't everything, it's the only thing," a quote that would become a central theme in his life. This ruthless approach to success and power would stay with Donald throughout his career, from his days as a young real estate mogul to his current political ventures.

In addition to these formative figures, Donald was also shaped by the teachings of Norman Vincent Peale, the pastor of the marble Collegiate Church in Manhattan. Peale's brand of Christianity was centered around the pursuit of personal success and prosperity. He taught his followers to focus on self-affirmation and the power of positive thinking, framing success as a divine right for those who believed in themselves. Donald, exposed to Peale's messages through his father's attendance at Peale's church, internalized this materialistic worldview, which viewed life as a series of individual battles rather than a communal effort. In this framework, the individual's ambition and success were paramount, and social justice or ethical considerations were secondary.

Beyond his father and mentors, Donald Trump also found guidance in figures who embodied the same cutthroat, manipulative tactics. One such figure was Roy Cohn, a lawyer known for his work with Senator Joe McCarthy and his ruthless approach to litigation. Cohn introduced Donald to the art of using threats, insults, and legal warfare to defeat opponents. Trump, who had long admired Cohn's ability to manipulate the press and power structures, adopted many of his tactics. Cohn's philosophy of "strike first, strike hardest" resonated with Donald, and their partnership would prove formative as Donald navigated the business world and later entered politics.

Donald Trump's rise to power was not just a product of his father's teachings but also a reflection of the cultural and political context of the time. Fred Trump, much like his son, used political donations not for ideological purposes but as a means to gain influence and access. Similarly, Donald's political maneuvering has been marked by opportunism, with his positions shifting to suit the moment. He has never been a man defined by consistent ideology but by the practical pursuit of power and influence, and his political success has often come at the expense of traditional norms and ethics.

The Trump family ethos, deeply rooted in the quest for success at all costs, led to a series of controversial actions and decisions that shaped Donald's career. His tendency to view life as a series of battles, in which strength and dominance are paramount, has permeated his approach to politics. This worldview has led him to embrace aggressive tactics in his campaigns, often alienating opponents and inviting extreme loyalty from supporters. In this, he is not simply following a personal philosophy but is enacting the legacy of a father who valued power, control, and self-interest above all else.

Today, as Donald Trump campaigns for the presidency, he continues to embody the lessons taught to him by his father, his military mentors, his religious influences, and his associates. His worldview remains one of intense self-interest, where rules are seen as obstacles to success rather than guides for ethical behavior. The influence of Fred Trump, who taught his son to win at any cost, lives on in every decision Donald makes, whether it's in business, politics, or his public persona. It is a legacy of ambition, ruthlessness, and a belief that victory is the only thing that matters.

In many ways, Donald Trump's journey has been an extension of Fred's values, magnified on the national stage. As a businessman, he learned to bend the rules for personal gain; as a politician, he adapted those same strategies to win support and achieve power. The lessons of his father have not just shaped Trump's career—they have defined it. In a world where personal success and domination are prized above all else, Donald Trump is the ultimate product of his father's ambition and influence.

Donald Trump's early years and his school record have been the subject of some public scrutiny. Here's a brief overview of his early education and associated controversies:

Cheating Allegations: In terms of cheating, there is no widely verified claim about Donald Trump cheating in school, but there are some stories that touch on his early education that might contribute to the perception of questionable behavior. His academic performance at NYMA was considered average, and he graduated from the school in 1964. Later, he attended Fordham University in New York City before transferring to the Wharton School of the University of Pennsylvania, where he earned a degree in economics in 1968.

Criticism of His College Years: Trump's college years have been the subject of media attention, but it's often been framed more in terms of his privilege and his relatively easy path to success compared to other students. There have been reports that Trump's father, Fred Trump, helped secure him a spot at Wharton, and his academic records have been kept private, making it difficult to assess his academic success accurately. Some critics have pointed to the lack of transparency surrounding his academic achievements. Trump's niece, Mary Trump, accused Donald Trump of paying someone to take the SAT in his place to help him get into Penn. Maryanne Trump Barry, Trump's sister and a former federal judge, says in tapes that the president got into the University of Pennsylvania through a proxy taking the entrance exam for him. At least half a dozen professors at the University of Pennsylvania's Wharton School have called on the University of Pennsylvania to investigate an allegation that Donald Trump paid another person to take the SAT exam on his behalf and was therefore fraudulently admitted to Wharton.

In addition, Fred Trump had several controversies regarding racism, sexism and even criminality in his business and personal dealings.

Racial Discrimination in Housing

One of the most significant allegations against Fred Trump came from his business practices in the 1960s. The U.S. Department of Justice (DOJ) sued Fred Trump and his company, the Trump Organization, for racial discrimination in 1973. The government accused the Trump Organization of systematically refusing to rent apartments to Black tenants in their buildings in Queens, Brooklyn, and Staten Island. This was part of a broader effort by the government to address racial discrimination in housing during that time.

The case, known as *United States v. The Trump Organization*, alleged that the company had intentionally told Black applicants that apartments were unavailable, while renting them to white tenants. Fred Trump and his company denied the accusations but ultimately settled the case, agreeing to a consent decree, which required them to take steps to ensure equal treatment of Black tenants. As part of the settlement, they also agreed to establish an affirmative marketing program and provide housing to Black renters.

The 1927 Ku Klux Klan Incident

Fred Trump's controversial history also includes a 1927 incident in which he was allegedly involved in a clash with police during a Klan rally in Queens. According to reports, Fred Trump was arrested (although not charged) during the unrest, which was sparked by a Ku Klux Klan rally in the area. While Fred Trump's role in the incident is debated, the fact that he was present at a rally organized by a notoriously racist group has raised questions about his attitudes toward race.

Fred Trump himself denied any involvement in the Klan rally, and he was never formally charged with any wrongdoing. However, this incident has been used by critics as evidence of his possible sympathies toward racist ideologies, given the context of the rally.

Refusal to Rent to Black Tenants

In addition to the 1973 DOJ case, Fred Trump's real estate empire faced other allegations of racial discrimination. In the 1960s and 1970s, some tenants and housing advocates claimed that Fred Trump had discriminatory practices in the way he managed his rental properties. These accusations often involved not just outright refusals to rent to Black families but also practices like limiting Black tenants to certain buildings or areas within a complex, effectively creating segregated living spaces. These actions were part of a broader pattern of discrimination in the real estate industry during that period, but the Trump family's involvement in these practices has been widely criticized.

Criminal Behavior

Allegations of Fraud and Deceptive Practices Fred Trump's business dealings also included various allegations of deceptive and unethical practices, some of which border on criminal behavior. One of the more notable accusations was that he used fraudulent methods to inflate the value of his real estate properties to secure loans from banks. Some legal analysts have pointed to his tactics as an example of "legal" manipulation or unethical business practices that could have crossed into criminality. While these actions were often covered up or resolved through settlements, they paint a picture of a man who wasn't afraid to bend the rules to get ahead.

Disputed Tax Practices Fred Trump's tax practices have also come under scrutiny. In a 2020 New York Times investigation into Donald Trump's tax returns, it was revealed that Fred Trump used a range of tactics to avoid paying taxes, including setting up trusts to pass wealth to his children without triggering inheritance taxes. It was suggested that Fred Trump may have engaged in fraudulent tax schemes that allowed him to pass on a fortune while minimizing the tax burden. While the details are complicated and may not have resulted in criminal charges, the actions outlined raised concerns about potential illegal tax avoidance.

Fred Trump also had an ultra-patriarchal view of women's place being in the home where they were tantamount to trophy wives or housemaids. He also did not hire or advance them for positions in his real estate business.

2. Early Business Ventures and Dealing with Family Wealth

Donald Trump inherited his family's real estate empire when Fred passed away in 1999. However, the road to that point was not without controversy. Early in his career, Donald faced significant challenges in gaining respect within the competitive New York real estate market. At first, his business ventures were funded largely by his father's wealth, and many of his early accomplishments were made possible by Fred's connections and financial backing. Critics argue that Donald's business ventures in the early years wouldn't have been possible without the financial head start from his father.

Throughout the years, Trump made headlines for his extravagant spending and brash personality. But behind the showmanship, there were a number of questionable business decisions, including taking risks with properties that were sometimes financially imprudent. He built an empire based on debt, often using loans from banks and investors that would sometimes fail to be repaid when his projects faltered. Trump's brashness in business often meant he was involved in numerous legal disputes, particularly regarding his dealings with contractors, partners, and creditors.

3. The Trump Organization: Controversial Practices and Legal Issues

Trump's business empire often found itself embroiled in lawsuits and disputes. One of the most notable controversies involved allegations of fraud in Trump University, which was accused of defrauding students by providing subpar real estate education programs. In 2016, Trump settled a class-action lawsuit for $25 million, although he did not admit wrongdoing. The Trump Organization also faced multiple lawsuits over unpaid debts and failure to adhere to contractual obligations with workers and contractors. Numerous labor disputes and accusations of mistreatment of workers followed Trump's real estate projects, where contractors often claimed they were not paid or were coerced into accepting lower wages than initially agreed.

Another major issue was his dealings with the casino industry. Trump's ventures into Atlantic City were marked by several bankruptcies and financial struggles. The Trump Taj Mahal, for instance, opened in 1990 but became a financial disaster, ultimately filing for bankruptcy in 1991. Despite his personal financial struggles, Trump continued to tout his own success, often branding himself as a master dealmaker, even as his business ventures teetered on the brink of collapse. Critics claim that Trump's aggressive tactics and disregard for the financial well-being of his businesses led to several of his ventures failing, with creditors left holding the bag.

4. The Role of Controversial Practices in Trump's Success

Trump's rise in business was also aided by his use of aggressive marketing and branding, even though many of his projects were often viewed as high-risk endeavors. While some of his properties became successful, particularly after he rebranded them with his name, Trump's persona as a businessman often overshadowed the failures and legal issues that plagued many of his ventures.

Additionally, Trump's family ties played a key role in his success, particularly as he maintained control over his company despite numerous failures and questionable deals. His father's influence ensured Trump never truly went without financial backing, and his connections in the world of New York politics and real estate provided him with advantages that many others did not have.

5. Roy Cohn's Influence on Trump's Political Style

Roy Cohn's influence on Donald Trump's career and political style is often described as one of the most formative relationships in Trump's rise to power. Their dynamic shaped many aspects of Trump's worldview, particularly when it came to power, strategy, and self-preservation. Cohn, a controversial and infamous lawyer, became Trump's mentor in the late 1970s when Trump was a young real estate developer trying to establish himself in New York City. Cohn's legacy as a ruthless attorney who thrived in the murky world of legal and political maneuvering had a profound impact on Trump's approach to business, politics, and personal interactions.

The Relationship Begins

Roy Cohn had gained national notoriety as the chief counsel to Senator Joseph McCarthy during the 1950s Red Scare, a period of intense anti-communist hysteria. He became known for his aggressive tactics, using his legal expertise to destroy the reputations of his opponents and manipulate the legal system. After McCarthy's downfall, Cohn continued to build a reputation as a cutthroat lawyer for powerful clients in New York. When Trump entered the public eye in the 1970s, he was already beginning to establish himself as a brash and ambitious figure in New York's real estate scene, and it was through this ambition that he crossed paths with Cohn.

Trump, at the time a young man with big dreams of expanding his father's real estate empire, sought Cohn's help when he found himself in legal trouble over his ambitious projects. Cohn, with his experience navigating the complex world of New York politics and law, became Trump's confidant and adviser. The two quickly formed a symbiotic relationship, with Cohn teaching Trump the "art of war" in the world of business, particularly how to use legal and media strategies to win battles, crush opponents, and project an image of invulnerability.

The "No Apologies" Philosophy

One of Cohn's key lessons to Trump was the importance of never apologizing, no matter the circumstances. Cohn's mantra was simple: "Never apologize, never explain." This attitude was grounded in the belief that showing any vulnerability or weakness would only invite further attack. Cohn was known for using personal attacks, smear campaigns, and aggressive legal tactics to destroy his adversaries, and he passed this approach on to Trump.

Trump would adopt this philosophy throughout his career, particularly in his dealings with the media and his political opponents. During his presidential campaign and beyond, Trump's refusal to back down from controversial statements, even when faced with widespread condemnation, became a defining trait of his persona. His approach was unapologetically confrontational, and his ability to weather scandal after scandal without offering a public apology mirrored Cohn's own tactics.

The Use of Media and Manipulation of Public Perception

Cohn was a master at manipulating the media and understanding the power of publicity. He recognized that controlling the narrative in the press could make or break a public figure, and he imparted this knowledge to Trump. Under Cohn's tutelage, Trump learned how to use the media to his advantage, turning even negative press into a tool to fuel his image as a tough, larger-than-life figure.

Trump, with his flair for self-promotion, used Cohn's lessons to create a persona that would make him a household name. Through carefully staged media appearances, interviews, and leaks, Trump turned himself into an icon of wealth and power. When the media criticized him, he used those attacks as evidence of his greatness, turning them into opportunities for self-aggrandizement. This strategy of turning controversy into an advantage would become a hallmark of Trump's public and political life.

The Cynical View of Law and Ethics

Cohn's influence also extended to Trump's approach to law and ethics. Cohn, who was disbarred in 1986 for unethical behavior, was known for bending or outright breaking the rules to achieve his goals. He believed that the law was just another tool to be used in the pursuit of power, and that legal ethics were secondary to the larger goal of winning.

Trump, who admired Cohn's success, adopted this cynical view of law and ethics, often treating legal boundaries as mere obstacles to be navigated rather than respected. This is evident in the numerous legal battles Trump has faced throughout his career, from his real estate deals to his political dealings. His approach to litigation has been marked by aggressive tactics, a disregard for legal precedent, and a willingness to push the limits of what is considered acceptable behavior.

Legacy in Trump's Political Style

As Trump transitioned from business magnate to political figure, the lessons he learned from Roy Cohn became even more evident. Trump's political style mirrored the aggressive, combative tactics Cohn had employed in his legal career. Whether it was his use of social media to attack opponents, his willingness to engage in personal insults, or his disregard for traditional political norms, Trump's approach was steeped in Cohn's influence.

Cohn's influence can also be seen in Trump's ability to connect with a certain segment of the American public that felt alienated by the political establishment. Cohn, who thrived in the underworld of political power, understood that there was a large audience that felt ignored or disrespected by the elites. Trump capitalized on this sense of resentment, positioning himself as the outsider who would take on the political establishment, much like Cohn had done for years with his legal clients.

The Personal Connection

Beyond the professional mentorship, Trump and Cohn shared a personal bond. Trump viewed Cohn not only as a mentor but also as a father figure. Cohn, who had no children of his own, seemed to relish the role of advising and guiding Trump. The two were often seen together in public, with Trump frequently praising Cohn's acumen and toughness. Even after Cohn's legal troubles and eventual disbarment, Trump continued to defend him, showing loyalty that seemed to transcend the boundaries of their professional relationship.

Conclusion

The influence of Roy Cohn on Donald Trump's career and political style is undeniable. Cohn's lessons in ruthlessness, media manipulation, and disregard for ethical constraints helped shape Trump into the figure he became in both business and politics. The relationship between the two men was a crucial chapter in Trump's rise, providing him with the tools he needed to navigate the cutthroat world of real estate and, eventually, the rough-and-tumble world of politics. Trump's approach to power, his unapologetic style, and his ability to manipulate the media all carry the fingerprints of Roy Cohn, who taught him that in the world of the powerful, the rules are made to be bent or broken—and it is the perception of strength that ultimately matters most.

Later, A.I. will delve into the relationships Trump had with sex slave human trafficker Jeffrey Epstein and has now with Nazi-saluting oligarch and potential coup architect Elon Musk.

PART II – Trump's Playbook: Project 2025

What will the term of Donald J. Trump's 2025-2029 Presidential Administration look like? What will it try to do, and what will that cost the American people both in their livelihoods and their very lives? And can he truly accomplish his 2024 campaign promises of bringing down inflation, bringing down grocery prices, deporting millions of immigrants and "making America great again?"

On the first day of his second term, Trump pardoned all January 6th Insurrectionist who stormed the Capital Building in Washington D.C. (January 6, 2021) and tried to overthrow the results of the 2020 U.S. Presidential Election that saw Joe Biden elected. Almost 1500 insurrectionists (all of them) were given a blanket pardon or had their sentences commuted by Trump's executive order, even if the insurrectionist was violent, hurt or killed cops, was part of domestic terrorist group(s) or had a prior record of criminal acts.

So what does Trump really have in mind moving forward in his second term in the Oval Office?

The answer may seem to not be easily accessible as Trump continues to write executive orders that are a random barrage of actions aimed against certain Biden administration orders and even laws created by Congress. A few of Trump's executive orders are already testing the constitutionality of his powers including trying to eliminate birthright citizenship, freezing financial funding to a vast array of government services that benefit all American citizens, firing inspector generals (who serve as federal watchdogs) without notice to Congress and letting an unelected and unconfirmed billionaire Elon Musk (who makes Nazi salutes) to have access to American payment systems of Social Security and Medicare among others.

There is however a playbook for Trump's "engineered incompetence" (as his detractors have termed his unpredictable and often vindictive governing style). It comes from the Heritage Foundation who wrote a 900+ page plan to serve as the blueprint for Trump and MAGA (Make America Great Again) Republicans to gut checks and balances of United States institutions and the very U.S. Constitution, which is the supreme law of the land.

The plan is called Project 2025.

Project 2025 is a plan created by Donald J. Trump and MAGA Republicans to reshape the U.S. government by removing checks and balances, consolidating power in the presidency, and enforcing a radical conservative agenda. Key elements include replacing civil servants with political appointees loyal to Trump, eliminating independent agencies, and pushing aggressive policies that affect personal freedoms and rights.

Main goals of Project 2025:

1. Restructure the government by dismantling federal agencies and empowering Trump's loyalists.
2. Roll back rights for women, LGBTQ+ people, and racial minorities, including banning abortion, restricting LGBTQ+ rights, and weakening anti-discrimination programs.
3. Attack middle-class and working families through cuts to social programs, public education, healthcare, and labor protections.
4. Increase risks to national security, the economy, and public safety by weakening oversight and undermining global alliances.

The project is driven by former Trump officials and supported by conservative organizations like the Heritage Foundation. It aims to give Trump sweeping control over the government from day one, weakening democratic safeguards and limiting personal freedoms.

Project 2025's sole purpose is claimed to be a conservative agenda to fix problems in the government, the economy and the collective America culture. Project 2025 declares it "paves the way for an effective conservative administration based on four pillars." Project 2025 is being organized by the Heritage Foundation. "The 2025 Presidential Transition Project is being organized by The Heritage Foundation and builds off Heritage's longstanding 'Mandate for Leadership,' which has been highly influential for presidential administrations since the Reagan era. Most recently, the Trump administration relied heavily on Heritage's 'Mandate' for policy guidance, embracing nearly two-thirds of Heritage's proposals within just one year in office." [About, Project 2025, accessed 6/24/24]

The executive summary mentions that Project 2025 was created by former Trump officials, right-wing operatives, and conservative backers, including Russ Vought, who is described as a Christian Nationalist and one of the project's authors. He served as Trump's Director of the Office of Management and Budget (OMB). Other members of the leadership team of Project 2025 include:

- Paul Dans: Director of the 2025 Presidential Transition Project. He was the former chief of staff at the Office of Personnel Management (OPM) during the Trump administration.
- Spencer Chretien: Associate director of the project, and formerly a special assistant to the president and associate director of Presidential Personnel.

The advisory board of Project 2025 consists of numerous conservative organizations and think tanks, but specific individuals who played a role in shaping or supporting the project aren't listed in the provided text by name beyond the leadership mentioned above.

It's also noted that many of the key players behind Project 2025 served in Trump's White House, and most are likely to serve in a potential future Trump administration. If you're looking for a more detailed breakdown, it might be available in further reports or public disclosures from the Heritage Foundation, which is behind organizing the project.

The Advisory Board for Project 2025 includes:

- 1792 Exchange

- **American Accountability Foundation**
- **AAPLOG**
- **Alabama Policy Institute**
- **Alliance Defending Freedom**
- **ACLJ Action**
- **American Commitment**
- **American Compass**
- **American Cornerstone Institute**
- **The American Conservative**
- **American Council of Trustees and Alumni**
- **American Family Association**
- **America First Legal (Stephen Miller's group recently asked to be removed from board though Stephen Miller is now in Trump's staff without Senate confirmation)**
- **American Juris Link**
- **ALEC**
- **The American Main Street Initiative**
- **American Moment**
- **American Principles Project**
- **The American Family Project**
- **Americans United for Life (recently dropped as a member of advisory board)**
- **AMAC Action**

- California Family Council
- Calvert Task Group
- Center for Equal Opportunity
- Center for Family & Human Rights
- Center for Immigration Studies
- Center for Renewing America
- Council for Citizens Against Government Waste
- The Claremont Institute
- Coalition for a Prosperous America
- Concerned Women for America
- Conservative Partnership Institute
- Defense of Freedom Institute
- Eagle Forum
- Ethics and Public Policy Center
- Family Policy Allowance
- Family Research Council
- Feds for Freedom
- First Liberty
- For America
- Forge Leadership Network
- Foundation for American Innovation
- Foundation for Government Accountability
- Freedom's Journal Institute
- The Frederick Douglas Foundation

• Heartland Institute

• Heritage Foundation

• MacArthur Society of West Point Graduates

• Hillsdale College

• Honest Elections Project

• Independent Women's Forum

• Institute for Education Reform

• Institute for Energy Research

• Institute for the American Worker

• Institute for Women's Health

• Intercollegiate Studies Institute

• James Madison Institute

• JCN

• Keystone Policy

• Leadership Institute

• Liberty University

• Mackinac Center for Public Policy (recently dropped as a member of advisory board)

• MRC

• Mississippi Center for Public Policy

• Moms for Liberty

• National Association of Scholars

• National Center for Public Policy Research

- Native Americans for Sovereignty & Preservation
- Noah Webster Educational Foundation
- OCPA
- Project 21 Black Leadership Network
- Pacific Research Institute
- Patrick Henry College
- Personnel Policy Operations
- Public Interest Legal Foundation
- Recovery for America Now Foundation
- Starrs
- Susan B. Anthony Pro-Life America
- Tea Party Patriots
- Texas Public Policy Foundation
- Teneo
- Turning Point USA
- Young America's Foundation

[Some of these may have changed since the production of this book, but they were in some part involved in Project 2025 creation.]

PART III – Project 2025's Components, Principles, Goals & Agenda

The Components of Project 2025 are as follows:

(1) Policy Agenda: The 920-page "Mandate for Leadership" provides a detailed policy and personnel blueprint for the next president to weaponize, restructure, or abolish every federal agency while also waging a "culture war" that includes cutting DEI (diversity, equity and inclusion) programs that champion the rights of BIPOC, LGBTQ+, disabled, military veterans, religious freedom seekers, etc. This wide-ranging "culture war" would even seek the absolute banning of pornography, as well as, suppressing protests with the use of violent military action (which is illegal though Trump would try to implement and exploit the Insurrection Act). There is even the proposal to deregulate child labor laws so teenagers would work "dangerous jobs" in "dangerous fields."

(2) Personnel Database: Collects thousands of resumes from MAGA loyalists to replace nonpartisan civil servants with political appointees that pass the MAGA loyalty test. Potential appointees and hires are asked when they had their "MAGA revelation." It is not an employee questionnaire, but rather a test of loyalty and a pledge of allegiance to Trump alone.

(3) Administration Academy: private training center to shape and prepare MAGA loyalists to take over government positions and politicize the civil service.

(4) 180-Days Playbook: Unpublished transition plan to immediately operationalize Project 2025, including executive actions and personnel changes to carry out Trump's promise of being "a dictator on Day 1." The executive orders that Trump has been signing originate from here.

The components of Project 2025 are based on the following principles:

Principles of Project 2025 has four main goals to shape America's future:

- **Restore the family as the centerpiece of American life and protect our children**
- **Dismantle the administrative state**
- **Defend our nation's sovereignty, border, and bounty against global threats**
- **Secure God-given individual rights to live freely**

These goals manifest themselves in the following agenda that has taken shape since Day 1 of Trump being sworn into the U.S. Presidential Office on January 20, 2025.

The summary of Policy Agenda of Project 2025 is as follows:

- **Guts Checks and Balances: Consolidated unchecked presidential power by purging the civil service, firing independent agency leaders, destroying entire agencies, and conditioning federal funding on political fealty. Project 2025 compromises the ability of every civil servant to uphold the law and forces them to serve the president at the expense of the public.**
- **Political Purge of the Civil Service: Imposes loyalty tests and purges thousands of independent civil servants so that the president can give control to thousands of political operatives, hand-picked for their unquestioned loyalty to the president regardless of qualifications, expertise, or commitment to constitutional duty.**
- **Seizes control of independent agencies: Fires independent agency leaders and forces independent agencies from the Federal Reserve to the FDA and the FCC to the CIA to**

prioritize the narrow interests of the president over their expertise, professional judgment and constitutional duty.

• Weaponizes the Justice System: Ends the independence of the DOJ, FBI and White House Counsel so that the president can turn them into the enforcement arms of the president,

including politicizing law enforcement, controlling prosecutions, targeting political opponents, and enforcing his agenda.

• Usurp the Power of the Purse: Alludes to a plan developed by the architects of Project 2025 to resurrect the Impoundment Act so that the president can unilaterally block congressionally-mandated spending. This has already begun with Trump's executive order to freeze all financial funding to all American citizens.

• Undermines Congressional Oversight: Evades the Senate's constitutional obligation to give advice and consent on the president's appointees and obstructs transparency, accountability and oversight.

Opponents of Project 2025 fear that the consequences of Trump and his administration enacting Project 2025 during his term will lead to:

Threatens Personal Freedoms and Rights

• Slashes Reproductive Freedom in Every State: Criminalizes medication abortions nationwide, bans abortion pills, creates an "abortion surveillance" regime to monitor all pregnancies to prosecute potential miscarriages, restricts birth control, and limits access to fertility services like IVF. It declares that the "Dobbs decision was just the beginning."

• Attacks LGBTQ+ Americans: Declares that anti-LGBTQ+ discrimination should be legal, rescinds nondiscrimination protections for LGBTQ+ Americans, undermines the rights of same-sex married couples, criminalizes gender nonconformity, bans transgender people from the military, weaponizes federal law to require states and private actors to discriminate against transgender people, and dismantles services for LGBTQ+ Americans wherever possible, including by mandating the federal government to officially adopt the stance that marriage is only between a man and woman.

- Undermines Racial Equality: Defunds services for underserved communities and emboldens racial discrimination in everything from housing to education to employment by systematically eliminating programs and policies that prevent discrimination or promote racial diversity, equity and inclusion. It further hinders efforts to address racial inequalities by prohibiting the collection of data related to race and ethnicity and by firing any federal employee that participated in a DEI program. In specific programs, it would absolve police departments under federal oversight for racial discrimination and wipe out the Office of Civil Rights responsible for protecting students and educators from discrimination.

- Rolls Back Women's Rights: On top of an unprecedented assault on reproductive freedom and sexual health, Project 2025 calls for the removal of "gender equality" or "gender equity" from every piece of federal legislation and reverses every policy, regulation and executive order aimed at promoting gender equality, including immediate dissolving the Gender Policy Council. Repeals any programs "subsidizing single motherhood" and eliminates services for working families, including overtime pay and Head Start. Rewrites Title IX and makes it harder to report sexual harassment and assault.

- Mass Detainments and Deportations: Carries out large-scale raids across the country to round up and detain millions of immigrants who have been living and working here for decades, including DREAMERs. Millions of people would be forced into detention camps near the border and deported en masse as part of the largest domestic deportation operation in American history.

• Deploys the Military Domestically to Quash Protests: The secret 180-Day Playbook reportedly includes a plan for Trump to invoke the Insurrection Act so that he can deploy the military to any state for domestic law enforcement and to put down protests and silence political opponents.

• Curtails Free Speech: Directs the DOJ to investigate and prosecute private actors involved in promoting equity and inclusion programs, rolls back online free speech protections, and contemplates denying access to the White House Press Corps.

• Outlaws Pornography and Bans Books: Seeks to outlaw pornography and imprison anyone involved in producing or distributing it. It implies that LGBTQ+ materials ar inherently pornographic, which is the pretext for book banning. Hence it seeks to register educators and librarians that share information related to transgender people as sex offenders. It also wants to shut any technology company that distributes the same.

Many believe these following reasons are why this will hurt the Middle Class and Working Families

• Ends Social Security and Medicare as we know them: Rips retirement and healthcare security from America's seniors. Although Project 2025 claims that Social Security and the Medicare Trust Fund "could not be covered here in depth," this conceals the fact that Project 2025's architects openly call for cutting Social Security, raising the retirement age, and turning Medicare into a privatized voucher program. Project 2025 also explicitly declares that it will make Medicare Advantage the default option, which would shift millions of senior citizens off of traditional Medicare and serve as a

backdoor for privatization. It also eliminates the Medicare Shared Savings Program.

- **Raises prescription drug prices: Undoes Medicare's new ability to negotiate prescription drug prices, eliminates the $35 monthly limit on insulin prices, and abolishes the cap on out of-pocket drug expenses for seniors. Trump has already signed an executive order to raise prices on a certain number of drugs.**

- **Guts Medicaid: Threatens to kick up to 18.5 million Americans off Medicaid by imposing draconian time limits or lifetime caps. Recommends further gutting Medicaid by turning it into an underfunded block grant or voucher program.**

- **Cripples Public Education: Completely eliminates the Department of Education, slashes public school funding, and redirects taxpayer dollars to religious private schools. This has not been enacted yet, but Trump spoke about doing this continually on the campaign trail in 2024 for his reelection.**

- **Eliminates Head Start: Eliminates the Head Start program, which provides preschool education for over 1 million children in low-income families. Project 2025 also rejects universal child care.**

- **Slashes food assistance: Guts SNAP and WIC while rejecting efforts to offer free universal school meals. This would have occurred during Trump's financial funding freeze he executive ordered in his first week of his 2024-2029 term. As of this writing, the courts have blocked the order being it violates the U.S. Constitution that gives "power of the purse" to Congress (which already had created the laws surrounding the financial funding during Biden's administration).**

- **Ends overtime pay: Allows employers to stop paying overtime and rolls back overtime protections.**

- **Raises taxes on the middle class and cuts taxes for wealthiest corporations and individuals: Restructures the tax** code to lower taxes for large corporations and wealthy individuals offset by tax increases on middle class families.

- Increases housing costs: Slashes funding for affordable housing, increases mortgage insurance premiums on FHA loans, and restricts eligibility for first-time homebuyer programs.

- Multiplies the costs of higher education and blocks student debt relief: Ends existing affordable student loan repayment plans, privatizes all student loans, and eliminates student loan forgiveness for teachers, nurses, law enforcement officers, public servants, and other eligible borrowers. It also calls for legislation to repeal any option for a student-friendly administration to cancel or forgive student loans.

- Rolls Back Labor Protections: Guts the capacity of the NLRB and EEOC to protect workers and calls for state waivers from key labor standards such as the minimum wage, overtime pay and right to organize. It also makes it harder for workers to organize and form unions, and it eviscerates child labor standards that protect children from hazardous jobs.

- Abandons our Veterans: Slashes federal jobs held by veterans and drastically limits the disability benefits for which veterans can qualify. It also bans transgender troops from service, revokes the VA's ability to provide abortion-related care, and slashes the number of general officers.

Furthermore, opponents of Project 2025 believe it puts Americans in Greater Danger being it would:

- Destabilize the Economy: Compromises the autonomy, professionalism and effectiveness of the civil service and independent agencies like the Federal Reserve and FCC which are responsible for protecting the stability of our economy through durable governing capabilities.

• Hurt Election Integrity: Eliminates CYBERCOM's role in protecting U.S. elections from cyberattacks, including from foreign actors.

• Damage Allies and Global Stability: Reorients alliances and pulls out of international organizations aimed at maintaining global peace and prosperity. Trump's tariffs and threats of tariffs have already run afoul of several allies like Colombia. His desire to annex Greenland and the Panama Canal have also strained or permanently undone relations with Denmark and Panama respectively.

• Undermine National Security: Purges the Intelligence Community of disfavored experts and civil servants to empower cronies loyal only to Trump, regardless of their qualifications in the intelligence community, which means the incompetent or corrupt may be carrying out matters of quintessential importance to national security.

Opponents also hold that Intelligence Community/National Security may be compromised being Project 2025:

• Undermines Public Safety: Makes it easier for people who want to commit violence to carry guns and harder for law enforcement to solve violent crimes.

• Threatens Consumer Safety: Curtails consumer protections and privatizes or politicize the FDA, FAA, TSA, and other agencies responsible for ensuring that our food, medicines, travel and products remain the safest in the world.

• Endangers Critical Health Research: Politicizes, restricts and ends NIH grants or investments in research labs and universities working to cure Alzheimer's, cancer, and other health issues.

• Limits Life-Saving Weather Alerts: Privatizes the weather service, which is responsible for preparing for extreme weather events such as heat waves, floods and wildfires. This will force Americans to pay for weather data and restrict access to life-saving weather alerts.

• Reverses Climate Action: Cuts scientific and climate research, dismantle environmental protections, and sell off federally protected public lands to the highest bidder. It also disbands the Environmental Protection Agency, undermining clean air and water protections.

Those reading this want to know what Trump will specifically do moving forward in his first 100 days in office, as well as, the next four years.

In summary, Project 2025 would:

- Arm the second Trump Administration with a comprehensive guide for gutting checks and balances
- in order to take over the federal government and grant the president unprecedented, unchecked power.
- Reinstate Trump's Schedule F Executive Order and purges the civil service of nonpartisan
- federal workers to replace them with thousands of political appointees hired based on their willingness to put loyalty to the president above their constitutional duty.
- Obstructs and intimidates apolitical and non-ideologically aligned civil servants to diminish their ability to fairly and independently serve the public interest.
- Empowers the president to fire independent agency leaders, defying Congress's considered
  choice to insulate these agencies against political interference.
- Compromises the functioning of our economy and public services by destroying entire agencies.
- Attacks public sector unions.

- **Politicizes grant-making by conditioning or withholding federal funding to States, local governments, nonprofits and private actors to force compliance with the Administration's extremist agenda.**
- **Abuses executive authority and the justice system.**
- **Obstructs transparency, oversight, and public participation.**
- **Undermines congressional oversight by evading the Senate's constitutional obligation to give advice and consent on the president's appointees and by blocking communication between agencies and oversight committees.**
- **Uses the powers of DOJ and independent regulatory agencies to investigate and prosecute organizations and individuals associated with DEI or ESG policies or programs.**
- **Use the power of granting/revoking security clearances to empower and neuter political agency personnel according to the Administration's wishes. Trump has already done this against Bolton, Fauci, General Milley who criticized him or tried to stop him from violating laws in his first term as U.S. President (2017-2021).**

PART IV – How Trump Would Take Reins of the Administrative State

Behind the scenes and Trump's bravado, Trump's staff and appointees would endeavor to "take the reins of the Administrative State."

Project 2025 advocates for centralizing power in the Oval Office under the guise of "deconstructing the centralized administrative state." It seeks to dismantle the

“Administrative State” by ensuring political appointees, loyal to the President, gain decision-making authority in the executive branch, eliminating reliance on “non-partisan experts” who may be seen as obstructing the President’s agenda. The project asserts that the next conservative President should prioritize dismantling this system, which it views as a corruption of the Constitution's legislative framework, where only Congress has the power to enact federal law.

The plan emphasizes filling government positions with political appointees loyal to the President, using them to “bend or break” the bureaucracy to align with the President's will. Success in achieving this goal would involve using the vast powers of the executive branch to reassert control, moving power away from Washington to local communities, states, and families.

Schedule F is central to Project 2025, reinstating Trump’s Executive Order 13957, which allows the President to install political operatives in the Executive Branch without expiration dates. This would force the conversion of apolitical federal employees into political appointments and grant the President the power to fire disloyal civil servants or those deemed non-ideologically compliant. Key goals of Schedule F include:

- Instating a limitless number of political appointees without expiration dates.
- Eroding protections for career federal employees, allowing the President to fire civil servants for political or ideological reasons.
- Removing merit-based hiring, favoring loyalty over expertise.

The impact of Schedule F includes politicizing the civil service, potentially firing as many as 500,000 federal workers, and undermining the impartial execution of government duties. It could also facilitate using government power to reward political allies or target opponents.

Additional plans to fire civil servants and replace them with loyalists include:

- A Heritage Foundation initiative to create a list of federal workers deemed resistant to the administration's agenda.
- Relocation of federal agencies to reduce the workforce or replace civil servants, as seen with the relocation of the National Institute of Food and Agriculture in 2019.
- A freeze on hiring for top career positions to prevent "burrowing-in" by outgoing political appointees.
- Shortening the disciplinary process for employees to make it easier to fire them.

Project 2025 also proposes granting political appointees the power to strip security clearances from civil servants and grant them to loyalists, which could undermine national security by placing unqualified individuals in sensitive positions. Additionally, the project seeks to use methods like biased ideological evaluations and IQ tests to select federal workers more aligned with a far-right agenda, further consolidating political control over the bureaucracy.

PART V – Defunding and Cutting Major Institutions like Education

Project 2025 advocates for reinstating Trump's Executive Orders (EOs) aimed at curbing union power within the federal government. Specifically, it calls for:

- EO 13836, which encouraged renegotiating all union collective bargaining agreements (CBAs) to ensure they align with management rights.
- EO 13837, which aimed to restrict union representatives from using official time to prepare or pursue grievances.
- EO 13839, which sought to limit labor grievances related to removals or performance appraisals and prioritize performance over seniority in retention decisions during reductions-in-force.

These orders were revoked by the Biden administration but are recommended for reinstatement. Additionally, Project 2025 suggests Congress reconsider whether public sector unions are appropriate, citing concerns over their compatibility with constitutional governance and good management practices.

Politicization of Grant-making

Project 2025 outlines several strategies to centralize grant-making power and align it with the President's agenda:

- It proposes that the Office of Information and Regulatory Affairs (OIRA) work with the Office of Management and Budget (OMB) to ensure grants follow centralized standards.

- It supports using the Office of Federal Procurement Policy (OFPP) to push back against "woke" corporate policies by influencing federal contracts and grants.
- A focus on reducing federal grant funding across government programs as a means to cut employee numbers and streamline operations.

FEMA Grants and Immigration Compliance

Project 2025 proposes conditioning FEMA grants on strict adherence to the administration's immigration policies, including:

- Certification of compliance with federal immigration laws, including honoring immigration detainers.
- Commitment to information-sharing with federal law enforcement agencies.

LGBTQ and Reproductive Health Language Removal

Project 2025 calls for the removal of LGBTQ and reproductive health references from all USAID materials and grants, advocating for a focus on women, children, and families instead. This includes eliminating terms like "gender equality" and removing all references to abortion and reproductive rights from agency publications, policies, and contracts.

International Aid Alignment to Anti-Abortion Policies

Project 2025 calls for aligning international aid funding with the administration's anti-abortion stance, implementing provisions that tie funding to organizations that uphold these values.

Teen Pregnancy and Abstinence Programs

The plan directs the Teen Pregnancy Prevention (TPP) and Personal Responsibility Education Program (PREP) to prioritize abstinence-only education programs in their grantmaking.

Block Granting Education Funding

Project 2025 recommends converting federal education funding into block grants to states, allowing families to use these funds for private school tuition. This would reduce federal control and increase state and local autonomy over education policy.

Cuts to Education Grants

The plan also suggests rolling back or eliminating education-related grant funding, particularly targeting regulatory programs like charter school priorities, civil rights data collection, student loan programs, and support for students with disabilities.

Investigation of Foreign Grants to Universities

Project 2025 proposes investigating universities that fail to comply with foreign-source reporting requirements and making them ineligible for federal financial assistance if they don’t meet these obligations.

Defunding Public Education

Project 2025 suggests eliminating competitive grant programs within the Department of Education and reducing funding for existing programs by 10%, saving approximately $8.8 billion.

This includes cutting federal support for K–12 education programs.

Defunding Climate-Related Programs

The project proposes defunding climate-related energy and infrastructure programs, particularly those managed by the Grid Deployment Office (GDO) and the Office of Clean Energy Demonstration (OCED). This includes rescinding unspent funds from the Infrastructure Investment and Jobs Act (IIJA) and halting subsidies for clean energy technology deployment.

In summary, Project 2025 seeks to drastically reshape federal grant-making, reduce public sector union influence, align international aid with conservative values, and cut federal spending across various sectors, particularly in education, climate-related programs, and public health.

PART VI – Project 2025's Plan for Your Wallet and the Economy

Inflation Reduction Act (IRA) and Grant Funding

Project 2025 seeks to repeal all programs within the Inflation Reduction Act (IRA) that provide grants for environmental science activities. The IRA's climate-related funding would be redirected, with an emphasis on providing grants to state regulatory entities rather than nonprofits, particularly in states that oppose climate change initiatives or support fossil fuel extraction.

Grants to Advocacy Groups
Project 2025 proposes halting all grants to environmental and climate advocacy groups, aiming to restrict funding to entities promoting these causes.

Environmental Justice and SCOTUS Ruling
In light of a potential Supreme Court ruling on race-based admissions, Project 2025 plans to halt environmental justice-related grants and pause actions under Title VI, pending a review to ensure consistency with SCOTUS decisions.

Climate Research Grants
Project 2025 would pause and review large grants to universities and institutions involved in climate research, placing a political appointee in charge to prioritize funding for projects that will have a tangible impact on the environment, without ideological or academic preferences.

NIH Grants and State-Controlled Scientific Research
Project 2025 suggests imposing term limits for top NIH career leaders and considers block granting NIH's budget to states to fund their own scientific research, potentially in partnership with other states for larger projects.

Sex Education Grants
Project 2025 would end support for evidence-based comprehensive sex education programs, shifting funding toward "Sexual Risk Avoidance" programs that emphasize

abstinence. It also calls for increased involvement of parents in these programs.

Family Planning and Anti-Abortion Funding

Project 2025 plans to amend the Title X family planning program to fund anti-abortion providers and defund organizations like Planned Parenthood. The focus would shift to fertility awareness and holistic family planning, while religious discrimination in grant selection would be eliminated.

Office of Justice Programs Grants

Project 2025 proposes using DOJ grants more effectively to support the President's prosecutorial agenda, particularly in local law enforcement, juvenile justice, and criminal justice research.

Transportation Grants

Project 2025 calls for shifting federal transportation grants to the states, allowing them to allocate funding for the transportation services that best suit their citizens' needs.

Economic Development Administration (EDA) Grants

Project 2025 recommends abolishing $350 million in Public Works Grants managed by the EDA and reallocating that funding to other federal priorities.

Small Business Administration (SBA) Grants

The plan proposes ending direct cash grants to small

businesses, highlighting the potential for fraud seen in previous SBA programs like the Economic Injury Disaster Loan (EIDL) Advance and Paycheck Protection Program (PPP), which saw billions in fraudulent payments.

Obstruction of Congressional Oversight
Project 2025 suggests manipulating the Federal Vacancies Reform Act to place political appointees in key federal agency leadership roles without Senate confirmation, thereby bypassing the oversight process.

Department of Homeland Security (DHS) Oversight
Project 2025 aims to limit congressional oversight of DHS by restricting communications between DHS and Congress, channeling all interactions through political appointees and minimizing outside influence.

Department of State Oversight
Similarly, Project 2025 recommends limiting congressional engagement with the State Department, directing all communications on certain issues through the White House, ensuring full coordination with political appointees.

In summary, Project 2025 seeks to reshape grant funding by redirecting resources from climate, environmental, and reproductive health programs to those aligned with conservative values. It proposes shifting control of scientific and educational grants to state governments, prioritizing programs with traditional family values, and limiting oversight by Congress through strategic use of political appointees.

PART VII – Placing Independent Agencies Under Direct Control of Trump

_Project 2025 proposes significant changes to the structure of the U.S. government, particularly in how independent agencies are controlled, the politicization of executive branches, and shifts in regulatory and policy enforcement. Here are key points:

1. **Independent Agencies Under Presidential Control:**
   - **Project 2025 suggests weakening the independence of federal agencies by placing them under direct presidential control. Independent commissions, like the FTC, have their leadership appointments but typically cannot be removed by the president. Project 2025 seeks to revisit the *Humphrey's Executor* ruling, which upheld the independence of multi-member agencies, arguing that the current system is constitutionally problematic and allowing for more presidential influence.**
   - **The proposal emphasizes using agencies like the FTC to further a conservative agenda, asserting that conservatives must leverage the power of government to prevent administrative expansion that conflicts with conservative goals.**

2. **Oversight of Independent Agencies:**
   - **Project 2025 advocates for the Office of Management and Budget's (OMB) Office of Information and Regulatory Affairs (OIRA) to oversee regulatory actions of independent**

agencies. This oversight would ensure compliance with executive orders, including reviews of significant regulatory actions, and extend to independent agencies like the Treasury Department, expanding the president's control over regulatory decisions.

3. **Examples of Specific Agencies:**
   - Federal Communications Commission (FCC): Project 2025 proposes eliminating the independent process and giving the president direct authority to influence FCC regulations, particularly against media outlets critical of the administration.
   - Federal Election Commission (FEC): The proposal suggests that a partisan shift in leadership could be used to control the enforcement of federal election laws.
   - Consumer Financial Protection Bureau (CFPB): Project 2025 aims to dismantle the CFPB, calling it unconstitutional. It proposes the return of consumer protection to the FTC and banking regulators, and suggests the president should be able to remove the CFPB director.

4. **Politicization of Executive Branch Agencies:**
   - The plan includes implementing the "unitary executive theory" (UET), which grants the president absolute control over the executive branch, including its agencies and military forces. This would allow the president to bypass Congress and the courts, consolidating power within the executive.
   - The proposal also advocates for the appointment of political loyalists in key positions, removing civil service protections for many federal employees to make them vulnerable to partisan shifts.

5. **Department-Specific Plans:**

- **National Security Council (NSC): The NSC would be empowered to fully implement presidential directives, enabling greater control over bureaucratic processes.**
- **Department of Education: Proposes eliminating the Department of Education and moving its Office of Civil Rights to the DOJ, shifting responsibility for civil rights enforcement to the DOJ. [Department of Education]**
- **Department of Justice (DOJ): Calls for limiting the independence of DOJ and FBI, potentially politicizing law enforcement to serve the administration's agenda.**
- **Department of Health and Human Services (HHS): Plans to reframe the department's mission, with a strong focus on supporting traditional family structures and tracking pregnancy data, including expanding reporting requirements on abortion and related health statistics. [Department of Health and Human Services]**
- **Department of Labor (DOL): Emphasizes filling DOL with political appointees to increase accountability to the president's agenda.**

6. National Oceanic and Atmospheric Administration (NOAA):

- **Dismantling and Downsizing: Project 2025 calls for the dismantling of NOAA, with its functions either eliminated, transferred to other agencies, privatized, or placed under state control. The proposal critiques NOAA for its role in the climate change debate, calling it a driver of "climate change alarm" and suggesting that its services could be more effectively provided commercially.**
- **Privatization: The plan proposes privatizing many of NOAA's functions to improve cost-effectiveness and quality.**

7. Department of Commerce:

- **Eliminating or Privatizing Functions: Project 2025 suggests reviewing the Department of Commerce to eliminate** overlapping programs, consolidate efforts, and possibly privatize some functions. The goal is to streamline operations and improve efficiency. [Department of Commerce, Project 2025]

- Census Bureau: Project 2025 recommends appointing strong political leadership at the Census Bureau to better align its mission with conservative principles and ensure efficient preparation for the 2030 Census. The administration should allocate additional political appointees to key positions.

8. Department of Homeland Security (DHS):

- Dismantling DHS: A key proposal is to dismantle DHS after 20 years of perceived disorganization, breaking it up into smaller, more mission-focused entities. The goal is to improve efficiency and reduce overhead. [Department of Homeland Security]
- Privatizing TSA: The Transportation Security Administration (TSA) would be privatized to reduce government involvement in airport security.
- Circumventing Senate Confirmations: The administration proposes placing political nominees into acting roles to bypass Senate confirmation processes, ensuring faster implementation of the Day One agenda.
- Restructuring Career Personnel: The plan also includes redistributing personnel within DHS to prioritize mission-critical functions and enhance law enforcement capacity.

9. Department of Veterans Affairs (VA):

- Undermining Whistleblower Protections: Project 2025 advocates for working with Congress to sunset the Office of Accountability and Whistleblower Protection (OAWP), arguing that it duplicates efforts already in place through other oversight bodies. The proposal seeks to streamline accountability processes and reduce redundancy.

10. **Environmental Protection Agency (EPA):**
   - **Dismantling Key Programs: Project 2025 aims to dismantle significant parts of the EPA, including the Office of Environmental Justice, and focus more on compliance than enforcement. The plan calls for slashing the EPA's budget and staff and halting climate-related investments under the IRA, particularly those that address climate change or strengthen community resilience.**
   - **Weaken Environmental Protections: A significant portion of the EPA's authority over air and water pollution would be reduced, limiting its ability to hold polluters accountable and undermining environmental justice initiatives.**

11. **Department of Housing and Urban Development (HUD):**
   - **Devolving Functions to States: Project 2025 proposes devolving many of HUD's functions to state and local governments and consolidating federal functions into other agencies. This includes transferring housing-related programs to agencies like the Small Business Administration and the Department of the Interior.**
   - **Politicizing Leadership: The plan calls for rapidly appointing political appointees to key HUD positions, converting career leadership into political roles, and using Senior Executive Service transfers to align the department with the administration's agenda.**

12. **Federal Reserve:**
   - **Overhauling the Fed: Project 2025 calls for drastically overhauling the Federal Reserve's role, limiting its mandate to focus solely on stable money. The proposal argues that the Fed's current broad powers over monetary policy and regulation are ineffective and politically overreaching.**

- **Abolishing the Economic Development Agency (EDA): The plan recommends abolishing the EDA and reallocating its funding to other federal grant programs, viewing it as an inefficient entity.**

These proposals reflect Project 2025's broader vision to reduce the size and scope of the federal government, privatize many public services, and centralize power within the executive branch, often at the expense of traditional government functions and oversight.

PART VIII - Changes to the Executive Office of the President

Here's an overview of the key recommendations made by Project 2025 related to the structure and functioning of the White House and various agencies:

1. **White House Counsel:**

- **Loyalty to the President and the Constitution: Project 2025 urges the next president to hire a White House Counsel who is loyal to both the president and the Constitution. The counsel should have broad legal expertise, including administrative and regulatory law, and a solid understanding of Congress and the political process. They emphasize loyalty over elite credentials even administering loyalty tests when candidates had their "MAGA revelation."**

2. **White House Media Space:**

- **Reexamining Media Space Allocation: Project 2025 suggests reconsidering the allocation of permanent media space on the White House campus. It argues that there is no legal entitlement for this space and that the next administration should balance media demands with space constraints.**

3. **Gender Policy Council:**

- **Abolishing the Gender Policy Council: The document calls for the abolition of the Gender Policy Council, which it claims promotes issues like abortion rights, comprehensive sexuality education, and gender-affirming care. Project 2025 advocates for rescinding relevant executive orders related to this council.**

4. **Reinstating Trump-era Executive Orders:**

- **Reviving Trump's Executive Orders: Project 2025 recommends reinstating various executive orders signed by President Trump that aimed to streamline the regulatory process, improve transparency, and reduce federal overreach. These include orders related to federalism, regulatory reviews, and fiscal responsibility.**

5. **Strengthening Regulatory Oversight:**

- **Information Quality and Privacy: The next president should strengthen implementation of the Information Quality Act, enforce the Paperwork Reduction Act, and ensure adherence to the Privacy Act. Project 2025 stresses the importance of regulatory cooperation to improve market conditions for America.**

6. **Administrative Pay-as-You-Go (PAYGO):**

- **Reintroducing Administrative PAYGO: Project 2025 advocates for reintroducing administrative PAYGO, a procedural requirement for agencies to ensure budget neutrality in discretionary spending, reinforcing fiscal responsibility.**

7. **Reining in the Administrative State:**
   - **Legislation to Reform Regulatory Processes: The next president should work with Congress to pass laws aimed at reducing the power of the administrative state. The document highlights several proposed reforms like the Regulatory Accountability Act, SMART Act, and REINS Act.**

8. **Maximizing the Congressional Review Act (CRA):**
   - **Undoing Midnight Regulations: Project 2025 recommends leveraging the CRA to undo late-stage regulatory actions from the previous administration. The proposed Midnight Rules Relief Act would enable Congress and the president to quickly repeal multiple regulatory actions.**

9. **Reshaping Climate Change Research:**
   - **Reshaping the US Global Change Research Program (USGCRP): The next president should issue an executive order to reshape USGCRP and related climate change programs, emphasizing the inclusion of diverse viewpoints and reexamining assessments prepared under the Biden administration.**

10. **Environmental Policy:**
    - **NEPA Regulation Reform: Project 2025 urges the president to direct the Council on Environmental Quality (CEQ) to rewrite regulations implementing the National Environmental Policy**

Act (NEPA), building on the 2020 reforms to streamline processes and reduce judicial review.

11. Eliminating Social Cost of Carbon Working Group:

- **Ending the Interagency Working Group on SCC: The next president should eliminate the Interagency Working Group on the Social Cost of Carbon (SCC) and end its analysis, which is seen as problematic for regulatory actions.**

12. Empowering the National Security Council (NSC):

- **Strengthening the NSC's Power: Project 2025 recommends empowering a strong NSC with full presidential authority to drive the bureaucracy and ensure that national security priorities are executed. This includes restructuring the NSC's staff to align with the president's goals.**

13. Staffing the NSC:

- **Replacing NSA Personnel: The document proposes replacing non-aligned personnel at the National Security Agency (NSA) with political appointees and trusted career officials, prioritizing alignment with the president's priorities.**

14. Limiting NSC Staff:

- **Reducing NSC Staff Size: To ensure better control, Project 2025 suggests limiting the number of NSC staff and minimizing the use of detailees, consolidating the NSC's functions to improve effectiveness.**

15. Trade Policy Leadership:

- Empowering the U.S. Trade Representative (USTR): The document urges the next president to empower the USTR to lead trade policy efforts, ensuring a coordinated approach across the government.

16. Office of Science and Technology Policy (OSTP):
   - Increasing OSTP's Role in Policy: If agencies are manipulating science for political or institutional agendas, the president should elevate the OSTP's Director, formally or informally, to ensure science plays a more prominent role in policy discussions.
   - Project 2025's proposals suggest a strong focus on consolidating power in the executive branch, reducing regulatory oversight, and reshaping agencies and policies to align with the new administration's goals, particularly in areas related to governance, environmental policy, and national security.

PART IX - Other Democratic Guardrails in Peril

Here's a breakdown of some of the more controversial elements outlined in Project 2025 and associated proposals that are tied to broader promises made by Donald Trump and his supporters:

1. Seizing Congressional Power of the Purse:
   - Canceling Federal Programs and Funding: Project 2025 advocates for a strong presidential role in defunding specific federal programs, even if they have been approved by Congress. This is seen as a means of curbing what they view

as "unaccountable" federal spending that supports "left-wing" bureaucracies. Trump has expressed plans to assert more control over the federal budget than any president before, potentially diminishing the constitutional role of Congress in fiscal decisions. If implemented, this would dramatically alter the separation of powers and could spark legal battles over the extent of presidential control over spending.

2. Weakening the Press:
   - Media Space and Press Access: The Project suggests that the next administration reconsider the provision of permanent media space on the White House campus. The idea is that there's no legal entitlement to this space, and the administration should rethink the balance between media access and space limitations. This could make it harder for journalists to effectively cover the president's activities. Additionally, there's an emphasis on reducing media influence, including ending funding for the Corporation for Public Broadcasting (NPR, PBS) and Voice of America, and potentially using the FCC to target media outlets that are seen as political opponents.

3. Deploying the U.S. Military Domestically:
   - Use of Military for Domestic Law Enforcement: Project 2025's influence extends to discussions around invoking the Insurrection Act, which would allow the president to deploy the military domestically. This could be used to suppress protests, dissent, or enforce a harsh immigration agenda. There are reports that this has been part of the planning for a second Trump term. The use of military force on U.S. soil would represent a significant shift in the role of the military and could raise serious constitutional concerns, especially in relation to civil liberties and public protests.

4. **Draconian Immigration Agenda:**
   - **Deportation and National Guard Use: The idea of invoking the Insurrection Act is connected to plans to enforce a stringent immigration policy, potentially deploying military forces and National Guard units for domestic operations. Some reports have suggested that the Trump administration could create a "private red-state army" to conduct mass deportations in defiance of non-compliant states. This would be seen as a major constitutional challenge, as it could override state sovereignty and raise issues about due process for individuals facing deportation.**
5. **Neutering the Senate's Role in Confirming Nominees:**
   - **Bypassing Senate Confirmation: In an effort to avoid Senate resistance to key appointments, Project 2025 proposes using "acting" positions to install individuals in senior roles without going through the Senate confirmation process. This approach has been used before by Trump and could be expanded to quickly implement a desired agenda. The Federal Vacancies Reform Act would be leveraged to place nominees into positions while their Senate confirmations are pending. This strategy could undermine transparency and accountability, bypassing the constitutionally mandated checks and balances of the Senate's role in confirmations.**

In summary, the promises and proposals linked to Project 2025, combined with Trump's broader plans, indicate a significant shift in executive power and a challenge to the traditional separation of powers, media freedom, and civil liberties. The strategies outlined could have long-lasting impacts on the balance between the branches of government, the rights of individuals, and the functioning of U.S. democracy.

PART X - Justice System and Law Enforcement

These would be changes to the Department of Justice and Other Law Enforcement Agencies under Project 2025.

Limits DOJ's independence:

- Requires dismissal of investigations, prosecutions, or cases deemed contrary to "Administration policies."
- Calls for reexamination of communication limits between White House officials and DOJ, which are currently restricted to the White House Counsel's Office and the Attorney General's Office, to avoid interference in criminal investigations and other matters like the issuance of Office of Legal Counsel opinions.

Expands political appointees:

- Advocates for a vast increase in political appointees throughout DOJ and the FBI, especially in the Civil Rights Division, the FBI, and the Executive Office of Immigration Review (EOIR) to "promote the President's agenda."

Curtails FBI independence:

- Proposes eliminating the FBI's General Counsel's Office and calls for ending the 10-year term for FBI Director, making the Director dependent on Presidential approval, reversing a Nixon-era reform.
- Calls for restructuring the FBI under the supervision of the Assistant Attorney General for the Criminal Division.

Separation of powers claims:

- Seeks to overturn Humphrey's Executor (a key legal precedent that allows independent agencies to have directors who are not removable at will by the President), asserting it violates the Constitution's separation of powers.

Weaponizes the Civil Rights Division:

- Focuses on investigating and prosecuting states, local governments, institutions, and private employers with diversity, equity, and inclusion (DEI) policies, labeling them as discriminatory.
- Bars the use of disparate impact in discrimination cases under Title VI of the Civil Rights Act.

Abortion-related policies:

- Misinterprets the Comstock Act of 1873 to block abortion pills and medical equipment, effectively banning abortion nationwide.

Investigates election-related crimes:

- Moves election-related investigations (like voter registration fraud) from the Civil Rights Division to the Criminal Division.

Prohibits FBI from tackling misinformation:

- **Bans FBI from combating misinformation and disinformation by citizens unless tied to criminal activity.**

Limits Federal Election Commission authority:

- **Urges changes to the Federal Elections Campaign Act (FECA) to remove independent litigating authority from the Federal Election Commission (FEC) and place it under DOJ control.**

Focuses on prosecuting leaks:

- **Expands investigations into unauthorized leaks, including classified information, to the media.**

Eliminates consent decrees:

- **Ends "all existing consent decrees" deemed inconsistent with Administration policy, including those protecting abortion access, while prioritizing prosecution of vandalism against crisis pregnancy centers.**

Immigration policy enforcement:

- **Assists Immigration and Customs Enforcement (ICE) in obtaining immigrant information from sanctuary jurisdictions.**
- **Pursues legal action against local officials (e.g., District Attorneys) who refuse to prosecute certain criminal offenses.**

Deploys DOJ to secure the border:

- Works with military and National Guard to assist in border operations targeting drug cartels and violent crime.

Drug and death penalty policies:

- Expands prosecution of interstate drug activity, including simple possession of marijuana.
- Prioritizes expanding the death penalty in current and future cases.

Overview

Project 2025 outlines a plan to curb the independence of the DOJ in multiple areas, including curbing the influence of the FBI and Civil Rights Division, revisiting legal precedents like Humphrey's Executor, and using DOJ to advance a conservative political agenda. It proposes major shifts in how the DOJ handles everything from abortion laws to voter rights to border security and the death penalty, aiming to significantly reshape its policies and structures.

PART XI – Impact on Health and Human Services

Here's a breakdown of key points regarding policies related to abortion, LGBTQIA+ rights, Medicare, Medicaid, and other health-related issues:

Abortion:

- **Propagates the belief that life begins "from the moment of conception," which could ban abortion, certain birth control methods, and assisted fertility treatments like IVF.**

- **Claimed repeatedly that abortion is not healthcare and proposed removing the word "abortion" from all existing rules, regulations, contracts, grants, and legislation.**

- **Would reverse FDA approval of abortion pills.
Seeks to reverse FDA's decision to lift restrictions on medication abortion, making it more accessible.**

- **Would ban federal funding for individuals forced to travel for abortion care.**

- **Proposes prohibiting Planned Parenthood from receiving Medicaid funds and urged Congress to defund the organization.**

- **Would not enforce federal law requiring hospitals to provide emergency abortion care as mandated by the Emergency Medical Treatment and Labor Act.**

- **Opposes facilitating abortion for unaccompanied minors in the custody of the Office of Refugee Resettlement.**

- **Pushed for restoring exemptions to the contraceptive coverage benefit under the ACA.
Claims there is no authority to add male condoms to the ACA's contraceptive coverage benefit.**

- **Seeks to exclude specific contraceptive methods from ACA coverage through misinformation about contraception.**

- **Proposed replacing the Reproductive Healthcare Access Task Force with an anti-abortion task force.**

- Believes research using human embryonic stem cells should not be taxpayer-funded and would end the human embryonic stem cell registry.
  Seeks to ban "three-parent embryo creation and human cloning research."

LGBTQIA+:

- Urged the CDC to avoid collecting gender identity data, labeling it as "destructive dogma."

- Opposes covering gender-affirming surgery under Medicare or Medicaid.

- Supports faith-based adoption agencies' right to discriminate against same-sex couples.

- Advocated for legislation to protect adoption and foster care providers from discrimination charges if they refuse to work with same-sex couples.

- Claimed that children raised in same-sex marriages are more likely to be unstable.

Medicare:

- Proposes repealing the provision in the IRA allowing Medicare to negotiate prescription drug prices.

- Supports repealing the Inflation Reduction Act and the Medicare Shared Savings Program.

- Suggests making Medicare Advantage the default enrollment option.

- Recommends using AI to reduce waste, fraud, and abuse within the system.

Medicaid:

- **Proposed implementing time limits or lifetime caps on benefits to "disincentivize permanent dependence."**

- **Urged CMS to "incentivize personal responsibility" and promote consumer choice for Medicaid recipients.**

- **Proposes allowing families to use Medicaid funds for coverage outside of Medicaid.**

Head Start:

- **Advocates for eliminating the federal Head Start program and the entire Office of Head Start.**

Non-Traditional Families:

- **Seeks to repeal Biden administration policies perceived as subsidizing single motherhood, disincentivizing work, and penalizing marriage.**

- **Would prioritize funding home-based childcare over universal childcare.**

CDC:

- **Called the CDC the "most incompetent and arrogant agency" in the federal government.**

- **Opposes the CDC enforcing school vaccination or mask mandates in the future.**

- **Proposes decommissioning the CDC Foundation.**

- Wants to rescind the legal analysis that allowed a moratorium on evictions during the pandemic.

NIH:

- Recommends block-granting NIH's grant budget to states.
- Proposes decommissioning the NIH Foundation.

Affordable Care Act (ACA):

- Proposed removing restrictions on physician-owned hospitals imposed by the ACA.
- Proposes separating insurance plans between subsidized ACA exchange and nonsubsidized markets.

Abstinence:

- Recommended people delay sex to prevent pregnancy and avoid reliance on TANF.
- Argued that teaching sex education would "promote prostitution" and create a "funnel effect" for abortion facilities.

Health Care Reform (Project 2025)

Project 2025 advocates for a market-based, patient-centered health care system that empowers individuals to control their health care expenses and decisions. It emphasizes that economic principles suggest competition in the provision of goods and services, including health care, will lead to reduced costs and improved quality and options. The reform includes decentralizing regulation, with states as primary regulators of the medical profession, and limiting federal restrictions that could hinder providers' ability to innovate or effectively serve patients.

Head Start Program

Project 2025 calls for the elimination of the federal Head Start program, arguing it is riddled with abuse and has failed to deliver long-term academic value to children. The program's high budget ($11 billion) has been criticized for contributing to incidents of child abuse and neglect, and research shows it has little lasting positive impact. The report also calls for the rescinding of any COVID-19 related restrictions such as vaccine and mask requirements for Head Start centers.

Non-Traditional Families and Marriage Policies

Project 2025 promotes policies that prioritize the traditional nuclear family model (married mother, father, and children). It criticizes current federal policies under President Biden that promote LGBTQ+ equity and subsidize single motherhood, which Project 2025 claims disincentives work and penalizes marriage. The document advocates for policies that encourage stable, married families and stress the importance of fathers in child-rearing. It argues that homes with non-related boyfriends are among the most dangerous environments for children. Furthermore, it emphasizes that children should be raised by their biological parents, and adoption should be expedited in cases of unfit parents.

Childcare and Family Support

The reform stresses funding for home-based childcare rather than universal daycare. Project 2025 argues that children in daycare experience higher levels of anxiety and neglect, with poor developmental and educational outcomes. It suggests providing parents with funding to either stay at home or pay for in-home childcare, instead of promoting institutional daycare.

Centers for Disease Control and Prevention (CDC)

Project 2025 critiques the CDC as an ineffective and overreaching agency, particularly during the COVID-19 pandemic, where it was accused of misjudging vital aspects of the virus, such as transmissibility and necessary health measures. The document proposes splitting the CDC into two distinct entities—one focused on collecting epidemiological data and the other on public health policy—to prevent conflicts of interest. Project 2025 also insists that the CDC should not dictate prescriptive health policies, like mask and vaccination mandates, leaving such decisions to parents and medical providers.

National Institutes of Health (NIH)

Project 2025 suggests breaking the NIH's control over research by block granting its budget to states for their scientific research. It also calls for term limits for NIH leadership and the abolition of the NIH Foundation, which is criticized for being influenced by pharmaceutical companies. The NIH's focus on diversity quotas and the #MeToo movement under previous leadership is also criticized, calling for the dissolution of its Office of Equity, Diversity, and Inclusion.

Affordable Care Act (ACA)

Project 2025 proposes several reforms to the ACA, including removing restrictions on physician-owned hospitals, which were imposed by the ACA and currently limit consumer choice and competition. The document also calls for separating the subsidized ACA exchange market from the non-subsidized market, to relieve the latter from the ACA's costly regulations. Additionally, it advocates for clearer rules regarding direct primary care (DPC), specifically stating that DPC's fixed fee model should not be classified as insurance under Health Savings Accounts (HSAs).

Social and Sexual Health Policies

Project 2025 advocates for abstinence-based policies and emphasizes delaying sex to prevent unintended pregnancies. The reform promotes sexual risk avoidance (SRA) programs, ensuring that these receive equal funding and attention as other programs. It argues that teen pregnancy prevention programs should avoid promoting abortion or high-risk sexual behaviors. Additionally, the document insists that sex education should not serve as an advocacy tool for promoting sex or abortion, and calls for a reevaluation of curriculum and program funding, ensuring a focus on health rather than risk.

Funding for Marriage and Family Programs

Lastly, Project 2025 recommends using Healthy Marriage and Relationship Education (HMRE) funding to support high school programs focused on healthy relationships, marriage, and sexual risk avoidance. This approach prioritizes early prevention and education, which it claims is more cost-effective than intervening in already broken relationships.

This summary reflects Project 2025's proposed overhaul of health care, family policies, and public health agencies, promoting a shift toward market-driven solutions, traditional family structures, and an abstinence-based approach to social and sexual health.

PART XII – More Healthcare Reforms (Medicaid, Medicare, Etc.)

Project 2025 Summary: Key Recommendations and Proposals

1. Child Welfare Provider Inclusion Act:

- **Purpose: Protect faith-based adoption and foster care providers from discrimination based on religious beliefs about marriage.**
- **Rationale: Project 2025 argues that children raised in heterosexual marriages experience greater stability compared to those in same-sex marriages. It stresses that children need both a mother's love and a father's protection.**
- **Key Position: Faith-based organizations should be able to adhere to traditional, biblically-based definitions of marriage and family without facing penalties or discrimination.**

2. Ryan White HIV/AIDS Program:

- **Proposal: Withdraw guidance allowing Ryan White funds to pay for gender transition procedures or "gender-affirming care."**

- **Rationale: Project 2025 claims these procedures cause irreversible harm and should not be funded by the program.**

3. Family Structure and Antidiscrimination Policy:

- **Position: Project 2025 asserts that marriage between men and women is the "ideal, natural family structure" for raising children.**
- **Rationale: It argues that children should ideally be raised by the biological parents (a man and a woman) who conceived them, and opposes conflating sex with gender identity or sexual orientation in federal policies.**

4. Medicare and Medicaid Reforms:

- **Medicare:**
  - **Problem: Medicare and Medicaid are described as "runaway entitlements" that hinder medical innovation, encourage fraud, and have a financially unsustainable future.**
  - **Recommendations:**
    - **Repeal the Inflation Reduction Act and the Medicare Shared Savings Program.**
    - **Repeal the provision in the Inflation Reduction Act that allows Medicare to negotiate drug prices, which Project 2025 claims will reduce access to medications and limit new drug access.**
    - **Make Medicare Advantage the default enrollment option to foster private health insurance.**
    - **Reduce the government share in the catastrophic tier of Medicare.**
    - **Use artificial intelligence to detect and reduce waste, fraud, and abuse in Medicare.**
- **Medicaid:**

- Funding Reforms: Proposes more flexible and efficient financing mechanisms for Medicaid, such as block grants, per capita caps, or blended match rates.
- State Flexibility: Encourages states to adopt work incentives for able-bodied individuals and provides the flexibility to modify benefit structures to better serve vulnerable populations.
- Personal Responsibility: Suggests that Medicaid recipients have a stake in their healthcare and contribute to costs in a way that protects taxpayers.
- Private Insurance Option: Proposes a "personal option" allowing families to use Medicaid funds to obtain coverage outside the program.
- Targeted Reforms: Suggests introducing time limits or lifetime caps on benefits to prevent permanent dependence and reduce entitlement costs.

5. Broader Policy Goals:

- Social Issues: Recommends affirming traditional family structures, with an emphasis on male and female biological realities.
- Medicaid's Role: Advocates for personal choice and responsibility in Medicaid, including allowing families to use funds for private insurance and incentivizing healthier behaviors.

Criticism and Opposition

Critics, such as *Rolling Stone*, argue that proposals like making Medicare Advantage the default enrollment option would accelerate the privatization of Medicare and significantly benefit private insurers, which could undermine the traditional Medicare program and reduce seniors' healthcare choices.

Summary

Project 2025 is a comprehensive set of policy recommendations aimed at reshaping U.S. healthcare, child welfare, and family structure. It stresses the importance of faith-based services in child welfare, proposes significant reforms to Medicare and Medicaid, and advocates for policies that align with conservative views on family and social issues. Key reforms focus on reducing government intervention, promoting private sector involvement, and limiting the scope of entitlement programs.

PART XIII – Labor, Unemployment, Kids Working Dangerous Jobs & Other Labor Reforms

Labor Policies

- Political Appointments: Maximize the number of political appointees at the Department of Labor (DOL). This approach aims to improve the department's political accountability and reduce the number of career officials, as political appointees are more directly aligned with the administration's agenda. The hiring freeze for career officials is designed to curb costs and reduce bureaucracy.
- Diversity, Equity, and Inclusion (DEI) Programs: End all DEI initiatives, including CRT-based trainings and racial classification systems within DOL. The goal is to reverse what they see as a "left-wing" ideology that promotes race and gender-focused policies and practices in the workplace. This would eliminate racial classifications and training systems that emphasize diversity as a key factor in labor policy.

- **EEOC Data and Demographics: Ban the collection of EEO-1 data by the EEOC, which currently tracks racial and ethnic data about the workforce. The intention is to prevent such data from being used in legal claims based on disparate impact theories, which they argue can unfairly lead to racial quotas or discrimination against businesses that don't meet racial or ethnic diversity targets. Additionally, amend Title VII to make such data collection illegal in both public and private workplaces.**
- **Discrimination Based on Gender Identity and Sexual Orientation: Legalize discrimination based on sexual orientation, gender identity, and transgender status. They suggest restricting the application of the Bostock decision (which protects employees from discrimination based on sex, including sexual orientation and gender identity) and rescinding regulations that protect transgender workers in terms of hiring and firing decisions.**
- **Religious Exemptions: Provide religious organizations with clearer exemptions under Title VII, allowing them to make employment decisions based on religious beliefs without violating nondiscrimination laws. This also includes enacting policies that protect religious employers and ensure they can enforce workplace rules that align with their faith-based values.**

Spending Cuts and Budgetary Priorities

- **Budget Reductions: Reduce the Department of Labor's (DOL) budget to historical lows, setting the target at $10.9 billion, which mirrors the Trump administration's 2020 fiscal year request. The goal would be to shrink the government's role in labor issues. The reductions would primarily target the Employment and Training Administration (ETA), which handles workforce development programs, and shift focus away from less critical inspections, especially those that don't involve major safety issues.**

- **Inspection Focus: Change health and safety inspection priorities to focus exclusively on the most egregious offenders, shifting away from routine inspections that may be seen as burdensome or redundant. This would reduce the role of federal inspectors in smaller violations and push more responsibility to state and local governments.**

Union Policies

- **Weaken Labor Unions: There are multiple proposals to dismantle union power. This includes:**
  - **Firing the general counsels of the National Labor Relations Board (NLRB) and EEOC immediately upon taking office.**
  - **Reorienting NLRB's enforcement priorities to be less supportive of unions.**
  - **Raising the threshold for NLRB jurisdiction, limiting the reach of federal labor laws on smaller businesses.**
  - **Diluting union power through policies like eliminating the "contract bar rule" (which makes it harder for unions to be decertified during active contracts) and allowing the Office of Labor-Management Standards to investigate unions without the need for a formal complaint.**
- **Alternative to Unions: They suggest creating "Employee Involvement Organizations", which would be company-controlled groups positioned as an alternative to unions. These organizations would not have the same power as traditional unions and would primarily serve to dilute union membership and limit the reach of organized labor.**
- **H-2A Visa Program: Cap and phase out the H-2A visa program, which allows foreign workers to fill seasonal agricultural jobs. This could be seen as an effort to curb the flow of foreign labor into the U.S. agricultural sector.**
- **Public Sector Unions: Question the legitimacy of public-sector unions altogether, potentially limiting their ability to operate or negotiate on behalf of government employees.**

Regulatory and Executive Actions

- Eliminate OFCCP: The Office of Federal Contract Compliance Programs (OFCCP) enforces executive orders that require federal contractors to adhere to nondiscrimination policies. Project 2025 suggests eliminating OFCCP by rescinding Executive Order 11246, which would simplify compliance rules for federal contractors and reduce regulatory burdens. This would particularly impact the enforcement of nondiscrimination policies based on sexual orientation and gender identity, which are not explicitly required by Congress but are enforced via executive orders.
- Rescind Regulations: Rescind regulations that prohibit discrimination on the basis of sexual orientation, gender identity, and transgender status, particularly in the federal contracting sector, which could pave the way for a reversal of protections for LGBTQ+ employees under Title VII.

Anti-Abortion Labor Laws

- Employer Exemptions from Abortion Benefits: Project 2025 calls for Congress to pass a law that clarifies employers are not required to provide accommodations or benefits for abortion. It proposes a law mandating that employers who provide abortion benefits must also offer equal or greater benefits for pregnancy, childbirth, maternity, and adoption. The intent is to ensure that pro-life options for mothers are given equal weight in employee benefits packages. Additionally, the law would explicitly clarify that employers aren't required to accommodate or provide abortion-related benefits.
- State Abortion Restrictions and ERISA: The proposal emphasizes ERISA (Employee Retirement Income Security Act), arguing that it should not preempt state laws that

restrict abortion or surrogacy. Project 2025 advocates that ERISA should not override state efforts to limit abortion-related benefits provided through employer-sponsored health plans. This would empower states to impose restrictions without the federal law preempting them.

Gut the EEOC and NLRB

- **Leadership Changes: One of the main proposals involves replacing the EEOC (Equal Employment Opportunity Commission) and NLRB (National Labor Relations Board) general counsels immediately upon taking office. This would mirror actions taken by the Biden administration in removing the previous counsels and push forward an agenda that aligns with the new administration's vision.**
- **EEOC's Regulatory Role: The EEOC's role is questioned, with Project 2025 urging the agency to disclaim its regulatory authority under Title VII and related laws. The EEOC would be urged to focus on guidance and technical assistance instead of issuing regulations, as it currently does, which are often seen as attempts to introduce new policies beyond its legal authority.**
- **No Consent Decrees by EEOC: Project 2025 proposes that the EEOC should not have the power to enter into consent decrees that mandate actions from employers, which could extend beyond what the law requires. This would limit the EEOC's ability to enforce settlements and mandates on companies.**
- **NLRB's Jurisdiction: They also recommend raising the revenue threshold at which the NLRB can assert jurisdiction over employers, effectively limiting the scope of the NLRB's influence over smaller businesses and reducing its role in regulating labor practices for these entities.**

Changes to the Women's Bureau (DOL)

- Redirection of Research: The Women's Bureau at the Department of Labor (DOL) would see a shift in its research focus. Instead of focusing on politically charged findings, Project 2025 suggests that the Bureau should reinvest in empirical research on the challenges facing women in the workforce, particularly regarding the earnings gap between men and women. The goal is to focus on open inquiry to understand the true causes behind these issues.

Allow Children to Work Dangerous Jobs

- Amend Hazard-Order Regulations: Project 2025 proposes that the DOL amend its hazard-order regulations, which currently prevent young workers (under the age of 18) from working in certain dangerous jobs. With parental consent and appropriate training, young workers would be permitted to take on more dangerous roles, particularly in industries experiencing workforce shortages. This would allow minors to build skills and gain experience in dangerous fields earlier.

Exempt Small Businesses from Certain Regulations

- Regulation Exemptions for Small Businesses: Project 2025 advocates for the exemption of small businesses from certain burdensome regulations that may disproportionately affect them compared to larger companies. This includes pushing for discretion under the Regulatory Flexibility Act to remove small business regulatory burdens wherever possible.
- OSHA Fine Exemptions: The plan also proposes that small businesses, particularly those with first-time, non-willful violations, should be exempt from fines issued by the Occupational Safety and Health Administration (OSHA). This

would be a step toward easing regulatory pressure on smaller enterprises.

Apprenticeship Programs

- **Encouraging Religious Organizations in Apprenticeships: Project 2025 recommends that religious organizations be encouraged to participate in apprenticeship programs and other workforce development initiatives. It suggests that historically, religious organizations have played a key role in advancing workers' dignity and should be given more freedom to participate in these programs without facing regulatory barriers. The initiative also calls for greater support for faith-based organizations to help workers in ways beyond just employment, like through vocational training and benefits networks.**

Union-related Reforms

- **Investigation of Unions: The Office of Labor-Management Standards (OLMS) would be given the authority to investigate unions without a formal complaint, similarly to how employers can be investigated for malfeasance. This would ensure that unions are held to the same standards of transparency and accountability as employers.**
- **Card Check and Secret Ballots: Project 2025 proposes eliminating the "card check" system (which allows unions to be recognized if a majority of workers sign union cards) and insists on secret ballots exclusively for union recognition. This is aimed at increasing transparency and reducing potential coercion in unionization efforts.**
- **Decertification of Unions: It also calls for the elimination of the contract bar rule, which limits employees' ability to decertify their union until the union's contract has expired.**

This would allow employees to more easily remove unions they no longer wish to be represented by.

Limiting Immigration and Visa Programs

- Capping the H-2A and H-2B Visa Programs: Project 2025 suggests capping the H-2A visa program (agricultural workers) and the H-2B visa program (non-agricultural seasonal workers). It also proposes a gradual phasedown of these programs over the next 10-20 years, with the goal of reducing reliance on foreign labor and encouraging higher productivity in U.S. agriculture and other sectors.
- Mandatory U.S. Citizen Employment in Federal Contracts: In addition to capping visas, Project 2025 suggests that all new federal contracts should require that at least 70% of employees be U.S. citizens, increasing to 95% over the next decade.

Unemployment

- Staffing Flexibility for State Agencies: Project 2025 advocates for greater flexibility for state agencies that administer unemployment benefits and workforce development programs. It recommends that states should be able to hire contractors for these roles if contractors are seen as better suited for the job. The proposal aims to remove requirements that mandate using state employees and instead allows for a more competitive hiring process that emphasizes effectiveness.

  The plan also emphasizes that federal mandates should not require states to use either union or non-union labor specifically for these positions. The focus is on efficiency and performance rather than adhering to union membership or employment status.

To achieve this, Project 2025 recommends that the Department of Labor (DOL) re-promulgate the Trump-era staffing flexibility rule, and Congress should codify this flexibility into law.

Retirement Plans

- ERISA Investment Restrictions: Project 2025 proposes that the Department of Labor (DOL) prohibit the investment of ERISA (Employee Retirement Income Security Act) plans based on any factors that do not directly relate to investor risks and returns. This means eliminating any considerations for factors like Environmental, Social, and Governance (ESG) concerns or political motivations when making investment decisions for retirement plans.
- The goal is to return to the Trump-era approach, which emphasized a strict focus on financial returns and investor protection over any non-financial considerations. However, the plan also acknowledges that legitimate non-ESG factors, such as corporate governance, supply chain investments in America, and family-supporting jobs, could still be considered if they are seen as beneficial to investment outcomes.

Joint Employers

- Defining Joint Employers: Project 2025 urges Congress to pass legislation to codify the definition of "joint employers", which would be based on direct and immediate control of employees. This is a response to more expansive definitions in previous administrations that sought to include broader, more indirect control.
- Under the proposed definition, a joint employer would be someone who has a direct and immediate control over the terms and conditions of employment (such as hiring, firing, pay rates, etc.) of workers. This would aim to limit the scope

of joint employer liability and reduce burdens on employers who are seen as having a more indirect relationship with workers.

- The proposal calls for Congress to enact the Save Local Business Act, which would codify this narrower definition and reverse the broader rules that were proposed during the Obama and Biden administrations.

Summary

Project 2025 envisions a radical shift in U.S. labor policies, focusing on deregulation, reducing union influence, and cutting back on immigration while empowering states to impose stricter limitations on abortion and related benefits. These proposals are seen as a direct response to what Project 2025 views as the overreach of federal regulatory agencies and efforts to promote progressive causes in labor law. The plan emphasizes individual choice and freedom for businesses, particularly small businesses, from heavy government intervention. Project 2025 envisions a significant shift away from current progressive labor policies toward a model that emphasizes deregulation, fewer government interventions, and a restructured relationship between employers, employees, and the government. These proposals aim to reduce the federal government's role in labor issues, limit the influence of unions, and eliminate what are perceived as politically motivated diversity and inclusion initiatives.

PART XIV – Education and Childcare

The proposal outlined in Project 2025 aims to significantly reduce the role of the federal government in education by eliminating the Department of Education (DOE), restructuring existing education programs, and transferring many responsibilities to the states.

Key Proposals and Actions:

1. **Department of Education (DOE) Elimination:**
   - **Project 2025 calls for the elimination of the Department of Education entirely, emphasizing that families and students do not need federal intervention for education.**
   - **A Department of Education Reorganization Act (or Liquidating Authority Act) would be enacted to wind down the DOE and redistribute its programs to other government entities.**
   - **The DOE would be replaced by flexible, formula-driven block grants to states and districts, aiming to minimize red tape and promote local control over education decisions.**

2. **Funding and Grants:**
   - **A significant reduction in the number of federal programs managed by the Office of Elementary and Secondary Education (OESE), with many either transferred to other agencies or eliminated.**
   - **Title I funding for low-income districts would be transferred from the DOE to the Department of Health and Human Services (HHS), with funds administered as a block grant.**
   - **Over a 10-year period, federal spending on Title I and other education programs would be phased out, shifting the responsibility to state governments.**

3. **Changes to Specific Programs:**

- Title I programs would face a 10% reduction annually, with complete federal funding elimination within a decade.
- Head Start and the GEAR UP program, which assists low-income middle and high school students, would be eliminated.
- Indian education programs would be moved to the Bureau of Indian Education.
- The Public Service Loan Forgiveness program and student loan forgiveness for specific occupations would be terminated, while student loans and grants would be privatized.
- Special Education (IDEA) funding would be converted into no-strings-attached block grants, and earmarks for special institutions would be phased out.

4. Title IX and Civil Rights in Education:
   - Project 2025 calls for a rollback of Biden-era Title IX regulations and the reinstatement of rules from the Trump administration, defining "sex" as biological sex and narrowing the definition of sexual harassment.
   - Restrictions would be imposed on the teaching of Critical Race Theory and discussions around gender identity in schools.
   - A ban would be placed on public education employees or contractors using names or pronouns not in line with a student's biological sex without written consent, with exceptions based on religious or moral convictions.

5. Higher Education:
   - The Higher Education Act would be amended to prevent accreditation agencies from mandating diversity, equity, and inclusion (DEI) policies or imposing standards that conflict with religious institutions' beliefs.

- Action would be taken against college accreditors, particularly the American Bar Association (ABA), using antitrust laws.

6. State Control and Parental Rights:
   - State governments would gain greater control over how education is provided, including for low-income families, with parents able to use Title I resources for private education options, including tutoring.
   - Legal protections for parents would be strengthened, with court hearings mandated if parents claim federal policies undermine their rights to raise, educate, and care for their children.

Education Savings Accounts (ESAs):

- Proposal for ESAs: Project 2025 advocates for Education Savings Accounts (ESAs) as a means for parents to have control over their child's education funding. The goal is to allow parents to direct their child's share of funding (largely funded by state and local taxes) to a variety of educational options that suit the child's unique needs, including private schools, tutoring, or educational resources.
- The model is inspired by Milton Friedman's 1955 proposal, where education is publicly funded, but decisions are made by families. This approach aims to shift the decision-making power from government entities to parents, increasing flexibility and personal choice in education.

Student Loans and Federal Student Aid (FSA):

1. Privatization of Student Loans: Project 2025 advocates for privatizing student loans and grants, shifting away from

direct federal lending. It suggests that the federal government's role should be limited to acting as a guarantor of loans, rather than a direct lender. The goal is to introduce more market-driven accountability in the higher education borrowing process by allowing private lenders to offer subsidized and unsubsidized loans, backed by government guarantees.

2. **Reversal of 2010 Student Loan Federalization:** The proposal calls for reversing the student loan federalization of 2010, which centralized federal student loan programs. Instead, the goal would be to create a new government corporation responsible for managing student loan obligations with professional governance and management, rather than keeping them under the Department of Education.

3. **End Loan Forgiveness Programs:** Project 2025 seeks to eliminate loan forgiveness programs such as:
   - **Public Service Loan Forgiveness (PSLF):** This program, which prioritizes government and public sector employment, would be terminated, as Project 2025 views it as favoring public sector over private sector employment.
   - **Time-Based and Occupation-Based Forgiveness:** It calls for ending forgiveness programs based on time or occupation, estimating that this could save $370 billion in taxpayer money.
   - **Borrower Defense to Repayment and Closed School Loan Discharge:** Regulations that expand loan forgiveness for borrowers based on institutional fraud or school closures would be rescinded. The new administration would propose that loan forgiveness should only occur with clear and convincing evidence of fraud by the educational institution.

4. **Phase Out Income-Driven Repayment Plans (IDR):** Project 2025 proposes to phase out existing IDR plans and implement a new, streamlined IDR plan with an income exemption equal to the poverty line. Borrowers would be required to pay 10%

of income above this exemption. Importantly, the new plan would not include loan forgiveness.

5. Ending Federal Payment Pauses and Loan Forgiveness Abuses: The proposal criticizes the previous administration's use of student loan forgiveness and payment pauses for political purposes, asserting that these programs must end to prevent abuses and ensure borrowers are held accountable to repay loans.

6. Removal of DOE's Authority to Forgive Loans: Project 2025 advocates for the removal of the Department of Education's authority to forgive loans under borrower defense to repayment, instead limiting loan discharges to cases of proven fraud by educational institutions.

7. Eliminate PLUS Loan Program: The PLUS loan program, which provides loans to graduate students and parents of undergraduate students, would be eliminated, which is expected to generate $2.3 billion in savings.

Office of Special Education and Rehabilitative Services:

1. Conversion of IDEA Funding to Block Grants: Project 2025 suggests that most funding under the Individuals with Disabilities Education Act (IDEA) should be converted into a no-strings formula block grant. These grants would be directly distributed to local education agencies (LEAs) through the Health and Human Services' Administration for Community Living. The goal is to give local agencies more autonomy in managing funds allocated for students with disabilities.

2. **Transfer of Vocational Rehabilitation Grants: Project 2025 proposes transferring Vocational Rehabilitation Grants intended for Native American students to the Bureau of Indian Education. This would shift the responsibility for these programs to a more specialized agency focused on Indigenous education.**

3. **Phase Out Earmarks for Special Institutions: The proposal suggests phasing out earmarks for various special institutions, aligning with the original vision of IDEA, which aimed to reduce the reliance on such earmarked funds and prioritize flexibility.**

4. **Rescind Regulations on Race and Ethnicity in Disability Discipline: Project 2025 calls for the rescinding of regulations under Part B of IDEA that require states to consider race and ethnicity when identifying, placing, or disciplining students with disabilities. The regulation, effective since 2017, would be eliminated as part of the broader effort to streamline special education policies.**

5. **Rescind the Equity in IDEA Regulation: Project 2025 proposes rescinding the Equity in IDEA regulation entirely, with no replacement regulation required. The goal is to eliminate mandates that may add layers of complexity to special education policies.**

6. **Move IDEA Oversight to HHS and Allow Parental Control Over Funding: The proposal recommends that oversight of IDEA be**

moved to the Department of Health and Human Services (HHS). Additionally, it suggests revising IDEA so that a child's share of funding could be made available to parents, allowing them to choose where and how their child learns.

Office of Postsecondary Education:

1. Move or Eliminate Office of Postsecondary Education (OPE) Programs: Project 2025 advocates for eliminating or moving the programs under the Office of Postsecondary Education (OPE) to the Employment and Training Administration (ETA) at the Department of Labor. This would aim to streamline and refocus federal efforts related to postsecondary education and workforce development.

2. Block-Grant Funding to HBCUs and Tribally Controlled Colleges: The proposal suggests that funding for higher education institutions should be block-granted, with a focus on narrowing these grants to support Historically Black Colleges and Universities (HBCUs) and tribally controlled colleges. This change would direct resources to institutions that serve minority and Indigenous populations.

Charter Schools:

1. Rescind New Requirements for Charter Schools to Receive Federal Grants:
   Project 2025 advocates for the next administration to rescind new requirements imposed on charter schools by the Biden administration regarding federal grants. Specifically, Project 2025 criticizes the 2022 changes to the Charter School

Program (CSP), which added additional requirements and regulations to charter schools applying for grants. The goal is to reduce the federal government's footprint in the charter school sector and lessen federal restrictions on charter schools, restoring the original intent of the CSP, which was designed to encourage the growth of charter schools with less bureaucratic interference.

2. Educational Choice for Children Act:
   Project 2025 suggests that Congress could consider passing school choice legislation, such as the Educational Choice for Children Act. This legislation would create a federal scholarship tax credit aimed at incentivizing donors to contribute to nonprofit scholarship granting organizations (SGOs). Families eligible for the program could use these funds for various educational expenses, including private school tuition, tutoring, and instructional materials. The proposal reflects a broader push to give families more flexibility in choosing how and where their children are educated, with an emphasis on empowering parents with more control over educational funding.

Higher Education Act (HEA)

1. Eliminate the Negotiated Rulemaking Requirement:
   - Project 2025 calls for Congress to amend the Higher Education Act (HEA) to eliminate the negotiated rulemaking requirement, which currently requires the Department of Education (DOE) to engage in extensive negotiations with stakeholders before making regulatory changes. Instead, the administration suggests that public hearings could serve as an alternative method for gathering input.

2. **Prohibit Accreditation Agencies from Mandating Diversity, Equity, and Inclusion (DEI) Policies:**
    - Project 2025 urges an amendment to the HEA that would prohibit accreditation agencies from using their Title IV gatekeeper role (which controls access to federal financial aid) to enforce policies that mandate the adoption of diversity, equity, and inclusion (DEI) programs at educational institutions. This reflects a push to remove DEI as a criterion for accreditation.

3. **Protect Religious Institutions from Accreditation Mandates:**
    - The proposal suggests protecting faith-based institutions by prohibiting accreditation agencies from imposing standards or criteria that conflict with the religious mission or beliefs of the institution. This includes ensuring that agencies cannot intrude upon the governance of religiously controlled colleges and universities.

4. **Encourage New Accreditation Agencies and Combat Abuse of Power:**
    - Project 2025 advocates for new accreditors to be established, especially to counteract what they see as abuses of power by existing accrediting bodies. The next Secretary of Education should refuse to recognize accrediting agencies that impose non-educational, ideological requirements on colleges, and should support the creation of alternative accrediting bodies.

5. **Antitrust Action Against College Accreditors:**
    - The President is urged to issue an executive order pursuing antitrust action against college accreditors, with particular focus on organizations like the American Bar Association

(ABA). This move is aimed at curbing what Project 2025 sees as monopolistic or overly regulatory practices within the accreditation system.

National Education Association (NEA)

6. Rescind NEA's Congressional Charter:
   - Project 2025 recommends that Congress rescind the National Education Association's (NEA) congressional charter, arguing that the NEA represents a radical special interest group that supports left-of-center policies and policymakers. The proposal claims this would help eliminate the false impression that federal taxpayers support the political activities of the NEA.

7. Investigate NEA's Use of Federal Funds:
   - Project 2025 calls for hearings to be conducted to investigate how much federal taxpayer money has been used by the NEA to support political causes that favor a particular political party. The aim would be to assess whether the NEA's actions are in line with its original purpose and whether taxpayer dollars are being used appropriately.

Federal Workforce Requirements

1. Elimination of College Degree Requirements for Federal Jobs:
   - Project 2025 proposes that the next president should issue an executive order stating that a college degree will not be required for any federal job unless the job specifically demands it. This proposal aims to minimize the reliance on bachelor's degrees as a hiring requirement across federal agencies and seeks to increase access to federal employment

for individuals who may have relevant skills or experience but lack formal higher education credentials.

College Preparedness

2. Eliminate the GEAR-UP Program:

- Project 2025 advocates for the elimination of the GEAR-UP program (Gaining Early Awareness and Readiness for Undergraduate Programs), arguing that it is not the federal government's responsibility to fund initiatives that create a pipeline from high school to college. The proposal suggests that GEAR-UP's functions should be taken over by private organizations or handled at the state and local levels, where policymakers may be better positioned to address college preparedness based on local needs.

About GEAR-UP:

- GEAR-UP is a discretionary grant program that provides funding to states and partnerships to improve college readiness for low-income students in high-poverty middle and high schools. The program provides services to students starting no later than the 7th grade and offers college scholarships to low-income students. The grants aim to increase the number of low-income students who are prepared to enter and succeed in postsecondary education.

Data Collection:

1. Move National Commission for Education Statistics to the Census Bureau: Project 2025 calls for the National Commission for Education Statistics (NCES) to be transferred to the Census Bureau within the Department of Commerce.

This move would align education statistics with broader demographic and economic data collection efforts. The proposal also includes the possibility of shifting some education-related research to the National Science Foundation for better governance and productivity.

2. Make Student Data Available by Family Structure: The proposal suggests that the Department of Education (or the agency responsible for data collection) should make student data available by family structure, enhancing public access to insights about how family dynamics may influence educational outcomes. The data would be included in tools like the Data Explorer and could be overseen by the Census Bureau, which would consolidate federal data collection efforts.

Summary

In essence, Project 2025 proposes a significant overhaul of the federal education system by eliminating the Department of Education, shifting control to states, reducing federal funding and oversight, and reinforcing parental rights. The overarching theme of the reforms is to reduce federal control over both K-12 education and student loans, increase market-driven accountability, and shift decision-making powers to families and private lenders. These proposals aim to limit government spending on education and loans, reduce federal involvement in higher education funding, and end various student loan forgiveness programs. It emphasizes local control, the elimination of federal programs deemed ineffective, and a shift toward privatization of student loans and grants. The proposals also seek to limit federal involvement in social issues like gender identity and Critical Race Theory in schools.

PART XV – Civil Rights / Gender Identity/ Sexual Orientation

Project 2025 calls for a series of reforms to civil rights policies and regulations under the Department of Education (DOE), particularly concerning Title IX, Title VI, and the Office for Civil Rights (OCR). These reforms involve significant shifts in the enforcement of civil rights protections, particularly regarding gender identity, school discipline, and protections for women in athletics. Here's a summary of the key proposals:

Office for Civil Rights (OCR)

1. **Move OCR to the Department of Justice (DOJ):** Project 2025 recommends that the Office for Civil Rights be relocated from the Department of Education to the Department of Justice (DOJ). The goal is to streamline enforcement of civil rights protections through the DOJ and the federal courts, where OCR would focus solely on enforcing these protections via litigation.

Title IX and Gender Identity

2. **Rescind Changes to Title IX Regarding Gender Identity:** Project 2025 urges the next administration to rescind changes made under the Biden administration, specifically the inclusion of a "nonbinary" sex category in the Civil Rights Data Collection (CRDC). The administration would issue a new CRDC to collect data that aligns more closely with OCR's statutory enforcement authority.

3. Restore Betsy DeVos's 2020 Title IX Regulations:
   Project 2025 proposes to restore the 2020 Title IX regulations under Betsy DeVos, which defined "sex" as biological sex at birth. These regulations narrowed the definition of sexual harassment and introduced live hearings with cross-examination for sexual misconduct investigations. They would replace the Biden administration's proposed changes, which Project 2025 criticizes for redefining "sex" to include sexual orientation and gender identity.

4. Due Process and Faith-Based Protections:
   Project 2025 suggests amending Title IX to include stronger due process protections for accused individuals and strengthened protections for faith-based educational institutions, programs, and activities.

5. Restore Women's Rights and Due Process:
   The proposal emphasizes restoring the rights of women and girls in athletics and education while reinforcing due process protections for individuals accused of sexual misconduct, which Project 2025 claims were undermined by the Biden administration's changes.

6. Abandon Redefinition of "Sex":
   Project 2025 calls for the next administration to reject the redefinition of "sex" to include sexual orientation and gender identity in Title IX regulations across all departments, arguing there is no scientific or legal basis for this shift.

7. Drop Ongoing Investigations Based on Gender Identity:
   Project 2025 urges that all ongoing Title IX investigations that

rely on the redefined concept of "sex" (as gender identity or sexual orientation) should be dropped, and affected schools should be allowed to rescind policies adopted under pressure from the previous administration.

8. Prevent Federal Agencies from Withholding Services:
The proposal would prevent federal agencies like the USDA from withholding services from state or federal agencies (including schools) that choose not to adopt the new gender identity definitions under Title IX.

Title VI – School Discipline and Disparate Impact

9. Review Title VI Cases on Disparate Impact:
Project 2025 calls for a comprehensive review of all Title VI cases (which concern racial discrimination) to determine how many of these cases are based on disparate impact (policies or practices that disproportionately affect one racial group over others). This review would assess how to align with the Trump administration's policies.

10. Joint Enforcement Guidance with DOJ:
The next administration should issue joint enforcement guidance with the DOJ stating that Title VI cases will not be investigated if they are based solely on allegations of disparate impact.

11. Clarify Title VI and Disparate Impact Standard:
Project 2025 recommends a new regulation clarifying that

Title VI of the Civil Rights Act does not include a disparate impact standard. This is aimed at ensuring that federal enforcement does not impose undue restrictions on school discipline or other policies.

12. Avoid Federal Meddling on School Discipline:
    Project 2025 calls for sweeping action to prevent the disparate impact standard from becoming a pretext for excessive federal involvement in school discipline and other local education policies. This would protect schools from potential federal overreach and protect the integrity of the Civil Rights Act in its original form.

Critical Race Theory (CRT)

1. Legislation to Prevent Discrimination:
    - Project 2025 urges lawmakers to pass legislation to prevent critical race theory (CRT) from spreading discrimination. The proposed legislation would ensure that no individual is punished or given benefits based on skin color, and that students and teachers are not required to believe anyone is guilty or responsible for the actions of others based on race or ethnicity. This reflects a push to curb CRT's influence in K–12 education.
    -

Parental Rights

2. Schools Serve Parents, Not the Other Way Around:
    - Project 2025 asserts that schools serve parents, not the other way around. The proposal advocates for universal school choice and states that any school boards, unions, or individuals who oppose parental authority should be cut off

from federal funding. This position emphasizes the role of parents as primary educators.

3. **Federal Parents' Bill of Rights:**
   - **Project 2025 calls for a federal Parents' Bill of Rights to restore parental rights as a top-tier right. The legislation would give families the right to a fair hearing in court when the government interferes with their ability to raise, educate, and care for their children. It would also require federal agencies to meet a strict scrutiny standard when enacting policies that impact parental rights.**

4. **Support for Existing Parental Rights Legislation:**
   - **The proposal supports various congressional bills that have been introduced to enhance parental rights, including the Empowering Parents Act (H.R. 8767), Parents' Bill of Rights Act (H.R. 6056), and constitutional amendments related to parental rights (H.J. Res. 99). Project 2025 recommends reviewing and prioritizing such bills to complement existing state-level parental rights laws in states like Georgia, Florida, and Virginia.**

Gender Identity

5. **Radical Gender Ideology and Its Impact on School-Aged Children:**
   - **Project 2025 warns that radical gender ideology is having a devastating effect on school-aged children, particularly young girls. It suggests that the next administration should address these concerns and take steps to protect children from what they view as harmful practices related to gender identity.**

6. Model Legislation on Names and Pronouns:
   - Project 2025 calls for the next administration to work with Congress to pass model legislation that would require K–12 schools to adhere to the following:
     - Names: Public education employees and contractors cannot address students by any name other than the one listed on the student's birth certificate without written permission from the student's parents or guardians.
     - Pronouns: Public education employees and contractors cannot use pronouns addressing a student that differ from the student's biological sex without written permission from the student's parents or guardians.
     - Religious or Moral Exemptions: Public institutions would be prohibited from requiring education employees or contractors to use a pronoun that does not match a person's biological sex if it contradicts the employee's religious or moral convictions.

PART XVI - Agriculture and Nutrition

Here's a summary of the proposals from Project 2025 related to food stamps, crop insurance, meat inspections, federal dietary guidelines, and GMO labeling:

1. Work Requirements for Food Stamps

- Proposal: Project 2025 recommends implementing work requirements for food stamp recipients to address what it describes as a "family crisis" in the U.S. This initiative is part of a broader conservative agenda aimed at using federal power to address issues like family breakdown and to push for policy changes that encourage self-sufficiency.

- Goal: The work requirement would limit access to food assistance for individuals who do not meet certain work-related criteria.

2. Increased Costs for Crop Insurance

- Proposal: Project 2025 proposes increasing the cost of crop insurance for farmers by reducing the premium subsidy that taxpayers cover. Currently, taxpayers fund about 60% of the premium costs for federal crop insurance policies.
- Goal: The aim is to reduce government spending on these subsidies, but it could result in higher costs for farmers purchasing crop insurance.

3. Elimination of Some Federal Meat Inspections

- Proposal: Project 2025 suggests eliminating certain federal meat inspections, allowing state-inspected meat to be sold across state lines without federal approval. This would reduce federal involvement in the meat inspection process.
- Goal: The proposal seeks to reduce regulatory barriers for the sale of meat products, potentially increasing competition and lowering costs. However, it could raise concerns about food safety and consumer health.

4. Repealing Federal Dietary Guidelines

- Proposal: Project 2025 advocates for the repeal of the Federal Dietary Guidelines, emphasizing that nutrition and dietary choices should be a personal responsibility, not dictated by the government.
- Goal: The proposal is grounded in the belief that individuals should be free to make their own dietary choices without

federal government interference, particularly in the form of dietary recommendations.

5. Elimination of GMO Labeling Requirements

- Proposal: Project 2025 calls for the elimination of the requirement to label genetically modified organisms (GMOs) in food products. It also supports maintaining federal preemption, allowing states to handle labeling laws on their own, but making sure voluntary labeling remains an option.
- Goal: The aim is to reduce regulatory burdens on food producers while giving consumers the option to choose whether or not to seek out GMO information voluntarily.

<u>Summary of Key Objectives</u>

- Food Stamps: Introduce work requirements for food stamp recipients to address issues of family structure and self-sufficiency. In other words, this is the end of welfare.
- Crop Insurance: Increase the cost of crop insurance by reducing government subsidies for premiums.
- Meat Inspections: Allow state-inspected meat to be sold interstate without federal oversight.
- Dietary Guidelines: Repeal the Federal Dietary Guidelines to promote personal responsibility for nutrition choices.
- GMO Labeling: Eliminate mandatory GMO labeling while allowing voluntary labeling and maintaining federal preemption over state laws.

These proposals reflect a conservative approach aimed at reducing government regulations and lowering federal involvement in agricultural and dietary policies. This could have wide-reaching and long-range health effects on

American citizens. It would also lead to environmental issues that are responsible for climate change.

PART XVII - Homeland Security

Here's a summary of the Department of Homeland Security (DHS) and immigration proposals outlined in Project 2025:

Department of Homeland Security (DHS)

1. **Dismantling DHS:**
   - **Proposal: The next president is recommended to dismantle the Department of Homeland Security due to its disjointed structure and inefficiency. The different missions of the DHS components are seen as conflicting, and breaking it up could reduce government overhead and create more focused missions.**

2. **Privatization:**
   - **TSA: The Transportation Security Administration (TSA) would be privatized, eliminating the need for federal management of airport security.**
   - **FEMA's National Flood Insurance Program: Privatize FEMA's National Flood Insurance Program, shifting much of the responsibility for disaster preparedness and response to state and local governments.**

3. **Restructure and Eliminate Agencies:**

- **FEMA, CISA, U.S. Coast Guard: FEMA could be moved to the Department of the Interior (DOI) or Department of Transportation (DOT), and CISA (Cybersecurity) would be shifted to DOT. The U.S. Coast Guard (USCG) could be moved to DOJ or DOD.**
- **Secret Service: Split into two parts: the protective mission moved to DOJ, and financial enforcement to Treasury.**
- **Office of Intelligence and Analysis: Would be eliminated, with its functions taken over by the FBI and other intelligence agencies.**

4. Eliminate Programs & Unions:

- **DHS Grant Programs: Most DHS grant programs would be eliminated to reduce spending.**
- **Unions: DHS unions would be removed for national security reasons.**

5. Immigration and Border Control:

- **Agency Mergers: CBP, ICE, USCIS, and other immigration agencies would merge into a standalone border and immigration agency at the Cabinet level.**
- **CBP and ICE Consolidation: If a full merger isn't achieved, CBP and ICE would be consolidated into a Border Security and Immigration Agency (BSIA).**
- **Increase Detention Space: ICE would see a significant increase in detention space, with 100,000 beds for detainees.**

6. Miscellaneous Proposals:

- **Border Patrol: Expand the horseback-mounted border patrol to increase border security presence.**
- **Secret Service and ICE Personnel: Recommendations to transfer personnel and adjust responsibilities, ensuring more efficient operations.**

Immigration Proposals

1. Visa Programs:
   - H-2B Visas: Do not increase the cap on H-2B (seasonal, non-agricultural) worker visas and end the H-2 eligible country list.
   - T and U Visas: Eliminate T and U visas for trafficking victims and crime victims, arguing that victimization should not be grounds for immigration benefits.
   - H-1B Visas: Transform the H-1B visa program into one aimed at attracting only the "best and brightest" foreign workers at the highest wages, to avoid disadvantaging U.S. workers.

2. Immigration Enforcement:
   - ICE Detention and Removal: Direct ICE to stop closing immigration cases, increase enforcement against criminal aliens, and remove all "sensitive zones" where ICE can't operate.
   - State and Local Enforcement: Allow state and local law enforcement to participate in immigration enforcement.
   - Credible Fear Standards: Raise the standards for credible fear asylum claims, particularly related to gang violence and domestic violence.

3. Eliminate Certain Immigration Programs:
   - Diversity Visa Lottery: Repeal the diversity visa lottery, ending random selection for immigration.
   - Chain Migration: End chain migration, focusing immigration on the nuclear family and promoting a merit-based system.
   - Temporary Protected Status (TPS): Repeal TPS designations and mandatory E-Verify for employers to verify work eligibility.

4. **Other Immigration Reforms:**
   - **Foreign Students: Limit visas for students from "enemy nations" and prioritize national security within educational exchange programs.**
   - **Merit-Based System: Push for a merit-based immigration system to prioritize high-skilled immigrants over family-based or lottery systems.**

5. **Immediate Deportation for Rejected Applicants:**
   Project 2025 proposes that applicants whose benefits or status adjudications are rejected should be required to leave the U.S. immediately until USCIS catches up with its case backlogs. Ordinary processes should resume only once these backlogs are resolved.

6. **Creation of Expulsion Authority:**
   The plan suggests establishing an authority similar to the Title 42 Public Health authority used during the COVID-19 pandemic, enabling the immediate expulsion of illegal aliens across the border when certain non-health conditions, such as loss of border control, are met.

7. **Repeal of Section 235 of the TVPRA:**
   The proposal advocates for the repeal of Section 235 of the William Wilberforce Trafficking Victims Protection Act, which grants immigration benefits to unaccompanied minors. This section is criticized for encouraging illegal border crossings and trafficking.

8. **End Flores Settlement Agreement:**
   Project 2025 calls for the end of the Flores Settlement Agreement, which sets standards for the detention of minors, and recommends establishing nationwide standards for family and unaccompanied minor detention. These standards should allow large-scale use of temporary facilities like tents.

9. **Increase in Credible Fear Standard for Asylum:**
   The proposal seeks to raise the standard for determining credible fear of persecution for asylum seekers, aligning it more closely with the asylum standard itself. Credibility determinations would also play a key role in the process.

10. **Exclusion of Gang and Domestic Violence as Grounds for Asylum:**
    Project 2025 suggests that gang violence and domestic violence should not qualify as grounds for asylum, proposing that Congress either eliminate or define the "particular social group" category more narrowly. This is in line with previous legal rulings, such as *Matter of A-B*.

11. **Halt Funding for NGOs Supporting Illegal Immigrants:**
    The plan recommends halting federal funds to non-governmental organizations (NGOs) that process and transport illegal immigrants, redirecting those funds towards securing the border, detaining aliens, and facilitating immigration court proceedings.

12. **Authorization of State and Local Law Enforcement in Immigration Actions:**
    Project 2025 suggests Congress authorize state and local law

enforcement to participate in immigration enforcement, in compliance with the *Arizona v. United States* decision.

13. Restrict Prosecutorial Discretion in Immigration Enforcement: The plan advocates for restricting the executive's ability to use prosecutorial discretion in immigration enforcement, removing what is seen as a "catch-all" excuse for limiting enforcement.

14. Mandatory Detention for Undocumented Immigrants: The proposal calls for eliminating discretionary language in Title 8 regarding the detention of aliens, ensuring that immigrants "shall" be detained rather than "may" be detained. This change aims to remove ambiguity and enforce Congressional intent.
15. Reinstatement of Asylum Cooperative Agreements with Northern Triangle Countries:
Project 2025 recommends using executive action to reinstate Asylum Cooperative Agreements with Northern Triangle countries to manage asylum requests more effectively.

16. Reimplementation of "Remain in Mexico" Protocols: The proposal suggests recommencing negotiations with Mexico to fully implement the "Remain in Mexico" Protocols, a policy requiring migrants to stay in Mexico while their asylum claims are processed.

17. Eliminate the Immigration Detention Ombudsman: Project 2025 calls for the elimination of the Office of the Immigration Detention Ombudsman, which was created under the Consolidated Appropriations Act of 2020.

18. Eliminate the Citizenship and Immigration Services Ombudsman:
The proposal recommends eliminating the Office of the Citizenship and Immigration Services Ombudsman (CISOMB), suggesting that its functions could be performed by other government bodies like the Office of the Inspector General (OIG) or the Government Accountability Office (GAO).

19. Deny Federal Loans to Students in Certain Schools:
Project 2025 suggests that the Department of Education deny federal student loans to students at schools that provide in-state tuition to undocumented immigrants.

20. Changes to FEMA Assistance Criteria:
The plan proposes raising the threshold for states and localities to receive public assistance from FEMA, citing that the current threshold has not kept pace with inflation, straining FEMA's resources.

21. Federal Government Cost-Sharing for Disaster Assistance:
Project 2025 suggests a new cost-sharing arrangement for federal disaster assistance, with the federal government covering 25% of the cost for small disasters and up to 75% for catastrophic disasters.

22. Wind Down National Flood Insurance Program (NFIP):
The proposal recommends phasing out the National Flood Insurance Program (NFIP) and transitioning to private

insurance, starting with lower-risk areas identified by the program.

23. Restructure FEMA Leadership:
Project 2025 suggests reducing the number of Senate-confirmed positions at FEMA from four to just one (the Administrator), eliminating the "springing Cabinet position," and streamlining FEMA's leadership to improve efficiency.

24. Revise U.S. Coast Guard Recruitment and Hiring:
The plan recommends re-vetting any promotions and hiring decisions made during the Biden Administration and re-onboarding U.S. Coast Guard personnel who were dismissed for refusing the COVID-19 vaccine. It also advocates for focusing on recruitment without emphasizing "wokeness" or diversity messaging.

Secret Service and TSA Reform in Project 2025

Key Proposals

- Visitor Logs for Presidential Residences: Project 2025 calls for the U.S. Secret Service (USSS) to maintain visitor logs for all facilities where the President works or resides, aiming to increase transparency regarding who might be influencing the highest levels of government. This proposal is a response to concerns over President Biden's use of his Delaware residence for work and the lack of visibility into visitors at such locations.
- Transfer USSS Agents to ICE: The proposal suggests that USSS agents stationed outside of Washington, D.C., be transferred

to Immigration and Customs Enforcement (ICE) field offices. This would involve closing non-essential USSS field offices and redistributing its personnel to enhance local law enforcement coordination and follow up on security threats, leaving the protective function of USSS focused on high-profile figures like the president and vice president.

- Deunionize TSA Workforce Until Privatization: Project 2025 proposes that the Transportation Security Administration (TSA) immediately deunionize its workforce and privatize aviation security screening. The plan aims to cut costs, by shifting to private-sector models seen in Canada and Europe, while keeping security levels intact. TSA would operate as a national security provider until privatized.
- Focus on EEO and Civil Liberties for DHS Civil Rights Office: Project 2025 suggests narrowing the role of the Civil Rights and Civil Liberties Officer at the Department of Homeland Security (DHS) to focus primarily on Equal Employment Opportunity (EEO) compliance and civil liberties issues. Investigations into civil rights violations would be forwarded to the Office of Inspector General (OIG) for review, with the Civil Rights officer only offering supportive information if the OIG declines to investigate.
- Security Failures on January 6th: The report highlights a specific security failure on January 6, 2021, when a pipe bomb was discovered outside the Democratic National Committee (DNC) building while Vice President-elect Kamala Harris was inside. It criticizes the U.S. Secret Service for not preventing this risk, suggesting that the agency's dual mission of protection and financial investigations distracts from its primary protective function, leading to preventable security lapses.
-

<u>Potential Consequences of These Proposals</u>

Impact on Secret Service Operations

- Visitor Logs and Transparency: Requiring USSS to keep visitor logs for any location where the President works or resides could potentially increase transparency, but it could also raise privacy concerns, particularly regarding the security of the President and high-profile government figures. This level of oversight might also be seen as politically motivated, potentially creating unnecessary friction between the Executive Branch and other stakeholders, such as state governments or private entities that host the President.
- Transfer of USSS Agents to ICE: Redirecting USSS agents to ICE field offices may streamline the agency's operations but could also lead to a reduction in security coverage for other high-risk locations outside of Washington, D.C. The increased emphasis on local law enforcement coordination might also overextend resources, making it harder to provide the specialized protection required for presidential protectees and other high-profile figures.
-

Impact on TSA and Aviation Security

- Privatization and Deunionization of TSA: The privatization of TSA could result in cost savings, but it might also lead to concerns about the accountability and training of private contractors handling sensitive security tasks. The push to deunionize the TSA workforce could lead to significant labor unrest and a reduction in worker protections, which may negatively affect morale and the effectiveness of security screenings. There could also be public safety risks if privatized entities cut corners to reduce costs.
- Potential for Reduced Oversight: Privatization might result in fragmented oversight of aviation security, where private contractors may not be as transparent or subject to the same rigorous regulations as federal employees. While cost-cutting measures may look appealing, they risk reducing the level of security and accountability at transportation hubs.

Impact on Civil Rights and Liberties

- **Narrowing the Civil Rights Role at DHS: Scaling back the DHS Civil Rights and Civil Liberties Officer's responsibilities could weaken the agency's ability to effectively address civil rights violations. Limiting the role to only EEO and civil liberties issues within DHS Headquarters might leave gaps in the handling of civil rights concerns across the entire department. Furthermore, relying on the OIG for follow-up on complaints could reduce public trust in the investigation process, particularly if the OIG fails to take action.**
- **Civil Liberties Investigations Handled by OIG: Shifting civil liberties complaints to the Office of Inspector General (OIG) for review might streamline oversight, but it could also create delays in addressing civil rights violations. The OIG typically handles broader issues of government misconduct, so this shift could leave some specific civil rights concerns unaddressed or under-prioritized.**

Impact on U.S. Secret Service's Protective Function

- Security Failures and the January 6th Insurrection: Highlighting security failures from January 6th as a "distraction" for the USSS suggests that the agency's resources may be spread too thin between its protective and investigative roles. This criticism may lead to calls for restructuring the agency, focusing more on its protective mission at the cost of other investigative responsibilities. However, such a focus on security could also potentially leave the Secret Service unprepared for handling new threats, as other intelligence or investigative duties may be ignored.

Overall Implications for National Security and Government Operations

- **Potential for Reduced Effectiveness:** The proposals to streamline and privatize various security and civil rights functions could lower operational costs but may also reduce the effectiveness of U.S. security agencies. Efforts to focus resources on specific missions (like protective functions for the Secret Service) may lead to other functions being under-resourced or poorly managed, creating gaps in national security.
- **Reduced Accountability and Oversight:** The deunionization of TSA and the narrowing of the Civil Rights Officer's role could lead to reduced worker protections and accountability. If these reforms lead to less oversight and fewer checks on power, it could result in policies that disproportionately affect vulnerable communities, undermine civil liberties, and reduce public confidence in government agencies tasked with safeguarding citizens.

These proposals reflect a strong push for streamlining government agencies, reducing federal spending, and tightening immigration enforcement. They aim to create a more efficient, privatized, and border-focused approach to both national security and immigration policy.

PART XVIII – Military, Defense and Foreign Adversaries

Military Recruitment and Personnel

- HIV-positive service members should be removed from service, and those with gender dysphoria should be expelled from military service.
- Reverse policies allowing transgender individuals to serve in the military, with a focus on eliminating public funding for transgender surgeries and abortions.
- Abolish diversity, equity, and inclusion (DEI) offices and staff, and eliminate Critical Race Theory (CRT) teaching.
- End tenure for academic professionals at military academies and eliminate all USMC law enforcement battalions.
- Improve recruiting by suspending the MHS Genesis system, which uses private medical records, leading to delays and unnecessary rejections.
- Ensure senior military officials focus on the readiness of the armed forces rather than social engineering.
- Reinstate service members discharged for refusing the COVID-19 vaccine, restoring their rank and providing back pay.

Abortion

- Ban the use of public funds to facilitate abortion for service members.

China

- Make defeating the Chinese Communist Party (CCP) the highest priority in foreign policy and defense.
- End economic engagement with China, emphasizing that past trade has enriched Chinese corporations while weakening U.S. industry.
- Outlaw TikTok, considering it a tool for Chinese espionage.
- Any university receiving Chinese money should lose accreditation and federal funding.

- Focus U.S. defense strategy on China, specifically prioritizing Taiwan's defense and countering China's Belt and Road Initiative globally.
- Reject moderate approaches to China, asserting that strategies accommodating China's rise have failed.
- Draft an Article X for China, with a comprehensive U.S. strategy focused on countering China's strategic culture and power.

Nuclear Capacity

- Expand and modernize the U.S. nuclear arsenal, focusing on restoring plutonium pit production and increasing investment in National Laboratories.
- Champion missile defense, reject claims that it destabilizes U.S. relations with Russia and China, and continue advancing missile defense technologies.

Alliances

- Transform NATO by making U.S. allies responsible for the majority of conventional forces, while the U.S. would primarily serve as a nuclear deterrent.
- Increase U.S. foreign military sales, enhancing global partnerships.

General Strategy

- Make irregular warfare a cornerstone of U.S. security strategy, integrating interagency and allied efforts.
- Employ offensive cyber capabilities against adversaries attacking U.S. infrastructure.

- **Prioritize a conventional force planning strategy to counter a Chinese invasion of Taiwan before allocating resources to other conflicts.**
- **Counter China's Belt and Road Initiative and engage globally to limit China's influence.**

Iran

- **Iran is at its weakest in its history, with internal protests and conflicts with regional neighbors signaling a weakening of the regime**

- **Criticism of the Iran Nuclear Deal: Project 2025 argues that the 2015 Iran nuclear deal (Joint Comprehensive Plan of Action) provided a monetary lifeline to the Iranian regime, which Project 2025 claims had been weakened by the 2009 Green Movement protests. The deal is said to have supported Iran's oppressive government, despite its people's desire for freedom.**
- **Promotion of Democracy: The plan advocates for U.S. support in encouraging a transition to a democratic government in Iran, focusing on human rights and leveraging economic and diplomatic tools to aid Iranian people's efforts for freedom.**
- **Policy on Iranian Ambitions: Project 2025 proposes preventing Iran from obtaining nuclear capabilities and expanding sanctions imposed during the Trump administration. It also emphasizes supporting regional partners, including Israel, to defend against Iran and its proxies like Hamas, Hezbollah, and Palestinian Islamic Jihad.**
- **Note on February 4, 2025, Trump said he would have all of Iran obliterated if they personally attacked him.**

Venezuela:

- **Response to Venezuela's Communism: The plan suggests uniting the Western Hemisphere against Venezuela's Communist regime and taking strong actions to aid its people. Project 2025 frames Venezuela as an underestimated but significant threat in the Southern Hemisphere, calling for robust actions to curtail its influence.**

North Korea:

- **Deterrence Strategy: Project 2025 asserts the need for continued deterrence against North Korea, which it views as a nuclear threat to the U.S. and its allies. It stresses the importance of preventing North Korea from threatening global stability through nuclear blackmail and violations of international agreements.**

Mexico:

- **National Security Threat: Project 2025 describes Mexico as a "national security disaster," citing its lack of sovereignty due to the power of criminal cartels. It calls for a new U.S. policy that supports Mexico regaining its sovereignty and tackling cartel violence.**

South and Central Asia:

- **Strengthening Ties with India: Project 2025 sees India as a crucial partner in countering China and securing a free Indo-Pacific. It advocates for bolstering U.S.-India relations and supporting India's role in the Quad, a cooperation framework that also includes Japan and Australia.**

- **Taliban and Pakistan: The plan dismisses the possibility of normal relations with the Taliban in Afghanistan or the military-led government of Pakistan, urging realistic and cautious engagement with these regimes.**

Nuclear Strategy:

- **Expansion of U.S. Nuclear Arsenal: Project 2025 calls for significant modernization and expansion of the U.S. nuclear arsenal to deter both Russia and China. This includes the restoration of plutonium pit production and investments in national laboratories to support the nuclear weapons program. The aim is to ensure readiness to address technological surprises and reinforce nuclear deterrence capabilities.**

Alliances and Defense:

- **Burden-Sharing and NATO: Project 2025 advocates for increased defense contributions from U.S. allies, particularly in the face of threats from China, Russia, Iran, and North Korea. It proposes transforming NATO so that European allies take on a larger role in conventional defense, with the U.S. focusing primarily on nuclear deterrence.**
- **Foreign Military Sales: The plan calls for a revival of U.S. foreign military sales, aiming to rebuild the defense industrial base and enhance interoperability with allies.**
- **Reorientation of U.S. Posture: Project 2025 suggests a fundamental shift in U.S. foreign policy to focus on honest assessments of global alliances, which could mark a significant change in post-Cold War foreign relations.**

Cybersecurity:

- **USCYBERCOM and Election Protection: Project 2025 advocates for ending U.S. Cyber Command's role in protecting U.S. elections from foreign interference, criticizing current efforts as politically biased.**

This strategic framework outlines a more aggressive stance on defense, a focus on curbing nuclear proliferation, and a redefined role for the U.S. in global security and alliances. Project 2025 does not go into great detail of accomplishing these endeavors along with geopolitical fallout would be if and when they would implement such policies.

PART XIX – International Relations & Tariff Warfare

Foreign Policy & International Relations:

1. **Withdrawal from Treaties and Alliances:**
   - **Exit from NATO and other international treaties.**
   - **Implement the "Remain in Mexico" policy.**
   - **Reduce or eliminate humanitarian aid and programs like the Export-Import Bank.**
   - **Cut economic ties with China and impose heavy tariffs.**

2. **Countries of Focus:**
   - **China, Iran, Venezuela, Russia, and North Korea are highlighted as top priorities, with an emphasis on national security and economic stability.**

3. **State Department Reform:**

- The next administration should overhaul the State Department, making it more efficient and aligned with the President's agenda.
- Emphasize political appointees in key roles, possibly bypassing Senate confirmation.
- Accept all resignations from ambassadors and reassess career diplomats.
- Freeze unratified treaties and review all international agreements to ensure they align with new policies.

4. Treaties and Agreements:

- Freeze implementation of unratified treaties.
- Review and potentially revoke agreements made without Senate approval, focusing on constitutional processes.
- Withdraw from treaties that have been under Senate consideration for more than 20 years.

5. Visa and Immigration:

- Enforce visa reciprocity policies, ensuring other nations provide equal access for U.S. nationals.
- Use Section 243(d) of the Immigration and Nationality Act to ensure countries accept the return of their nationals.
- Reevaluate visa programs like the Diversity Visa, F, and J visas for national security alignment.
- Temporarily reduce refugee admissions due to the ongoing border crisis.

6. Fentanyl Crisis:

- Collaborate with Mexico and other Western Hemisphere countries to tackle the fentanyl crisis, focusing on cartels and their connections to China.

7. **Manufacturing and Supply Chains:**
   - **Shift global manufacturing from places like China to countries in Central and South America to improve supply chains and reduce dependence on distant, unreliable sources.**

8. **Middle East:**
   - **Strengthen relations with Saudi Arabia and work to reduce Chinese influence there.**
   - **Build on the Abraham Accords and encourage more Arab nations to join.**
   - **Defund the Palestinian Authority and push back against Turkey's potential alliances with China or Russia.**
   - **Create a Middle East security pact involving Israel, Egypt, Gulf states, and possibly India.**

9. **Sub-Saharan Africa:**
   - **Focus on core diplomatic activities without promoting U.S.-specific social policies like abortion or LGBTQ rights, which may not be well-received in African countries.**

U.S. Engagement in Africa:

- **Social Policies: Avoid imposing U.S. social policies (e.g., abortion, LGBT rights) on African nations, focusing instead on security, economic, and human rights engagement.**
- **Economic Focus: Shift U.S. development assistance from humanitarian aid to fostering free market growth through incentivizing U.S. private sector engagement.**
- **Countering China: Prioritize combating China's influence in Africa, including developing public diplomacy to counter Chinese campaigns, supporting Somaliland statehood, and safeguarding sensitive communications networks.**

- Security Cooperation: Support African military and security operations through U.S. assistance and training programs.

U.S.-Europe Relations:

- Transatlantic Trade: Amplify trade relations with Europe to reduce European dependence on China.
- Review of Trade Agreements: Conduct a review of trade agreements between the U.S. and EU to ensure fair treatment of U.S. businesses, and prioritize the development of trade with the UK post-Brexit.
- New EU Allies: Strengthen diplomatic relations with Central European countries, especially those vulnerable to Russian aggression.

Arctic Policy:

- Strategic Importance: Acknowledge the Arctic's increasing geopolitical and economic importance due to melting ice, and safeguard national sovereignty.
- NATO Strategy: Develop an Arctic strategy within NATO to protect Arctic resources from excessive Russian use and ensure the region's stability.
- Economic Freedom: Promote economic freedom in the Arctic to maintain its stability and security.
- China's Arctic Ambitions: Work with Arctic nations, including Russia, to address China's growing influence and the so-called "Polar Silk-Road."
- Greenland Ties: Strengthen U.S.-Greenland relations by enhancing economic ties and establishing a year-round U.S. diplomatic presence in Nuuk.

International Organizations:

- Withdrawal from Ineffective Organizations: The U.S. should be prepared to withdraw from international organizations if they no longer serve U.S. interests, citing the WHO's failure during the COVID-19 pandemic as an example.
- Support for Effective Organizations: The U.S. should support international organizations that align with American goals but continually apply pressure to ensure their effectiveness.
- Praise for Trump Administration's Actions: The Trump administration's withdrawals from or cessation of funding to the WHO, UN Human Rights Council, UNESCO, and the UN Relief and Works Agency are praised as steps that served U.S. interests.
- Anti-Abortion Stance: The U.S. should not fund or promote abortion in international programs. The Geneva Consensus Declaration on Women's Health and Protection of the Family is highlighted as a successful initiative from the Trump administration.
- Geneva Consensus Declaration: The U.S. joined like-minded countries to promote anti-abortion stances internationally, arguing there is no international right to abortion and defending national sovereignty and family values.
- WHO's Human Rights Record: The WHO's actions during the pandemic are criticized for supporting the suppression of human rights, especially due to its relationship with China and other human rights abusers.

Foreign Aid:

- Consolidating Foreign Assistance: The Administrator of the U.S. Agency for International Development (USAID) should also oversee all foreign assistance. This dual role would align foreign assistance with U.S. foreign policy goals, as was done briefly under the George W. Bush administration.

State Department:

- **Freeze on Unratified Treaties: The U.S. should freeze efforts to implement unratified treaties (e.g., Paris Climate Agreement) under a Republican administration, pending a review.**
- **Paris Climate Agreement: Withdrawing from the Paris Agreement is proposed, though this could make the U.S. appear unreliable and worsen global emissions efforts, particularly benefiting countries like China.**

USAID:

- **Climate Change and Other Program Cuts: Project 2025 criticizes USAID's involvement in global climate change efforts, calling it "climate extremism." Proposed cuts would target climate programs, global gender equality, and anti-racism initiatives. Elon Musk has already illegally shut down USAID under the fictitious and unapproved DOGE.**

Export-Import Bank (EXIM):

- **Debate on the Future of EXIM: There is confusion within Project 2025 on whether to abolish the Export-Import Bank or retain it. The document includes contradictory viewpoints, signaling a lack of consensus on the role of the bank in supporting U.S. business interests abroad.**

PART XX - Environmental Deregulation and Defunding

Key Environmental Impact Proposals:

- **EPA's Capabilities Slashed:**
  - **Executive order on Day One to cut EPA budget, staff, and programs.**
  - **Will "pause and review" major environmental and public health protections, potentially weakening them.**

- **Disbanding Office of Environmental Justice and External Civil Rights (OEJECR):**
  - **The office tackles environmental justice, civil rights, and community engagement.**
  - **Consequences:**
    - **Increased costs: $2.7 billion in IRA funding for environmental justice efforts and over $50 million for community grants would be prevented.**
    - **Job losses: Over 200 jobs across 10 EPA regions would be lost.**
    - **Increased pollution: $3 billion in climate justice grants would be halted, reducing efforts to combat air pollution and climate change in vulnerable communities.**
    - **Kneecapping enforcement: Stopping technical assistance for environmental justice priorities, civil rights concerns, and enforcement of federal laws.**

- **Reversal of the Good Neighbor Plan (2022 Expansion):**
  - **The plan aimed to reduce ozone-forming emissions from power plants and industrial sources.**
  - **Impact:**
    - **130 million Americans (40% of U.S. population) live near non-power plant facilities.**
    - **Consequences:**
    - **Healthcare cost increase: Reversal would negate up to $15 billion in benefits by 2026.**
    - **Air pollution: Prevent 15% reduction in nitrogen oxide (NOx) emissions, increase 45,000 tons of NOx, 29,000 tons of sulfur**

dioxide, 1,000 tons of fine particles, and 16 million metric tons of CO_2 emissions.

- **Public health: Potentially 1,300 premature deaths, 2,300 hospital visits, and an additional 1.3 million asthma cases by 2026.**

- **Educational and economic impact: 430,000 school absence days, 25,000 lost workdays.**

- **Elimination of Office of Enforcement and Compliance Assistance (OECA):**
  - **OECA enforces compliance with environmental laws, including water system lead violations and chemical pollution.**
  - **Consequences:**
    - **Increased costs: Would prevent disbursement of $45 billion in nonfederal dollars for water infrastructure.**
    - **Job losses: 950 EPA enforcement jobs lost due to prior budget cuts; 200 of these jobs were restored in 2023 but would be undone.**
    - **Increased pollution: 21% increase in ozone emissions from electric power generation, 6 million metric tons of CO_2, and a 10% decrease in air quality in already polluted areas.**
    - **Public health and environmental justice risks: 2,250 additional lung cancer deaths, fewer inspections in environmentally overburdened communities, and loss of civil rights reviews.**

- **Rejection of EPA Office of Research and Development (ORD) Science Activities:**

- The proposal would restrict ORD's science-based activities and research unless authorized by Congress.
- Consequences:
- Economic and scientific harm: Over $800 million in cuts to scientific research on climate change and natural systems.
- Undermines climate change adaptation: Slows progress on carbon storage, wildlife conservation, and extreme weather preparedness.
- Public health and safety risks: Stopping efforts to release environmental and health data, hinder contamination remediation, and impede sustainable waste recovery initiatives.

Superfund Site Cleanups:

- **Project 2025's Proposal:**
  - Aims to cut costs for industry by undermining toxic and hazardous site cleanups, disproportionately affecting communities of color and low-income areas.
  - Consequences:
    - Increased health risks: Short-term, cost-cutting cleanups could lead to incomplete cleanups that jeopardize community health and safety.
    - Weakened regulations: Groundwater cleanup standards would be loosened, and PFAS (per- and polyfluoroalkyl substances), known as "forever chemicals" with serious health risks, may not be classified as hazardous substances under CERCLA.

Cumulative Impact Analysis & Climate and Equity Initiatives:

- **Proposal to End Cumulative Impact Analysis:**
  - Seeks to ban the use of cumulative impact analysis under the National Environmental Policy Act (NEPA) and limits judicial review of NEPA analysis.

- Aimed at giving project proponents more control over regulatory timelines.
- Consequences:
- Polluters over people: Industry would have a stronger voice than affected communities, especially vulnerable ones.
- Climate and health risks: These proposals would increase pollution, health threats, and economic insecurity, particularly in disadvantaged communities.
- Undermines equity: Disadvantaged communities, including communities of color and low-income areas, face higher pollution and climate threats. These changes could strip their right to clean air, water, and a healthy environment.

Interior: Fossil Fuel Expansion and Environmental Rollbacks:

- "Drill Everywhere" Proposal:
- A call for aggressive expansion of fossil fuel extraction, focusing on U.S. lands and the entire Western Hemisphere, including the Arctic.
- Willow Project:
- Suggests a near-doubling of drilling pads in the Arctic (expanding from three to five pads).
- Consequences:
- Climate setbacks: Expansion of oil and gas drilling across U.S. lands and the Western Hemisphere would worsen climate change.
- Health and safety risks: Increased drilling would exacerbate health and safety harms to marginalized communities, particularly Native and Indigenous peoples.
- Removal of Protections from Public Lands and Monuments:
- Proposes rolling back protections from national monuments, parks, and wildlife refuges, including Bears Ears, Grand Staircase, Katahdin Wood, Tongass National Forest, and the Arctic National Wildlife Refuge.
- Consequences:

- Environmental degradation: Removal of protections would harm critical ecosystems, wildlife, and public health.
- Public opposition: The American public overwhelmingly opposes removing protections for public lands and national monuments.

- Impact on Native American Issues:
  - The plan includes controversial recommendations such as facilitating fossil fuel development on Native lands and ending federal subsidies for electric vehicles.
  - Consequences:
    - Undermines sovereignty: The proposals would undermine Native American sovereignty and environmental protections.

Wildlife and Environmental Protection:

- Threats to Wildlife:
  - Proposes reinstating bearbaiting, a cruel practice to make black bears easier targets for trophy hunters.
  - Consequences:
    - Harm to wildlife: The plan would endanger wildlife and undermine efforts to protect endangered species.
    - Climate change impact: These actions would also hinder efforts to stop climate change and protect critical ecosystems.

Energy:

- End Grid Planning, Focus on Reliability:
  - Proposes halting federal grid planning and focusing only on reliability, sidelining efforts to expand renewable energy.
  - Consequences:
    - Missed opportunity for renewables: The proposal would block progress on improving transmission systems to facilitate renewable energy.

- Impact on power supply: The plan could stop efforts to address transmission bottlenecks, leading to power shortages.

- Defund Grid Deployment Office Programs:
  - Recommends defunding nearly $20 billion in grid modernization programs, including transmission facilitation.
  - Consequences:
    - Slows clean energy transition: Reducing federal involvement in grid modernization would slow the transition to clean energy and hinder efforts to combat climate change.
- Eliminate Loan Program Office:
  - Proposes eliminating the DOE's Loan Program Office (LPO), which funds innovation in clean energy technologies.
  - Consequences:
    - Job and economic losses: The LPO has created over 46,000 permanent jobs and supported technological advancements.
    - Hinders emissions reduction: The elimination of the LPO would reduce job creation and technological progress, slowing emissions reduction efforts.

PART XXI – Commerce, Treasury and Tax Reforms

Department of Commerce: Project 2025 Recommendations

1. Abolition of Advisory Committees
   - Review all advisory committees for necessity, abolishing those not required by statute.
   - Reconstitute remaining committees with experts to align with conservative principles.
   - End involvement of activists opposing conservative policy.

2. Dismantling NOAA

- Break up NOAA by downsizing, privatizing, or transferring many of its functions to other agencies or state control.
- Focus National Weather Service (NWS) on data-gathering, while privatizing forecasting services, citing private companies like AccuWeather as more reliable.
- Prioritize commercialization of weather technologies.
- Transfer survey operations to U.S. Coast Guard and U.S. Geological Survey for increased efficiency.
- Merge National Marine Fisheries Service and U.S. Fish and Wildlife Service.
- Harmonize Magnuson-Stevens Act with National Marine Sanctuaries Act to avoid conflicts with no-fishing zones.
- Withdraw support for the 30x30 Executive Order and the America the Beautiful Initiative, claiming they hinder commercial activities and promote offshore wind energy.
- Modify regulations under the Marine Mammal Protection Act and Endangered Species Act to avoid hindering fisheries and Native American subsistence.
- Reevaluate the need for environmental review processes, such as NEPA, on fisheries actions to reduce unnecessary delays.
- Disband much of the Office of Oceanic and Atmospheric Research (OAR) climate change research, which is viewed as contributing to climate alarmism.
- Reassign assets from the Office of Marine and Aviation Operations to other agencies.

3. Census Bureau

- Increase political appointees within the Census Bureau to guide preparations for the 2030 census.
- Review and potentially revise partnerships to improve census response, engaging conservative groups for better participation.
- Push for a citizenship question despite past opposition from the Supreme Court ruling.

- Reevaluate race and ethnicity questions to prevent political biases in data collection.
- Scrutinize and potentially remove non-crucial census questions.
- Abolish the Census Bureau's National Advisory Committee on Racial, Ethnic, and Other Populations.
- Disband non-statutory standing committees, including the Census Scientific Advisory Committee.

4. Economic Development Administration (EDA)

- Abolish the EDA and reallocate its funding to other federal grant programs.
- Build on the success of Opportunity Zones to incentivize private sector investment in distressed communities.

5. International Trade Administration

- Consider relocating the International Trade Administration (ITA) and parts of the Bureau of Industry and Security (BIS) to the U.S. Trade Representative (USTR).
- Ensure strict verification of imports, particularly from China, using advanced analytics and artificial intelligence to detect circumvention and prevent bad actors.
- Ensure senior policy positions are held by political appointees.
- Revive a China-specific non-market economy unit to oversee trade with China.
- Replace Top Market Reports with industry competitiveness assessments for strategic economic sectors.

6. Restructure Industry and Analysis (I&A)

- Restructure I&A to focus on ongoing supply chain analysis and identify vulnerabilities in critical sectors, like pharmaceuticals and semiconductors.

- Pursue strategic decoupling from China and strengthen U.S. defense industrial base and critical supply chains.
- Restrict or control exports of goods and technology to countries of concern unless justified by humanitarian or medical needs.

7. Foreign Trade and Investment

- Consolidate Global Markets and elevate Advocacy Center and SelectUSA to enhance large-scale export transactions and foreign direct investment (FDI), particularly for reshoring manufacturing and research & development.

8. USPTO (U.S. Patent and Trademark Office)

- Performance-based Organization: Transition USPTO into a performance-based organization under the Office of Management and Budget (OMB).
- Consolidation with NIST: Merge USPTO with the National Institute of Standards and Technology (NIST) to form a new U.S. Office of Patents, Trademarks, and Standards, while eliminating or relocating non-critical research functions.
- Opposition to IP Waivers for Cutting-edge Tech: Oppose granting intellectual property waivers, including for COVID-19 vaccines and other advanced technologies, through international agreements like the WTO's TRIPS.

9. Economic Development Administration (EDA)

- Grant Program Consolidation/Elimination: Consolidate or eliminate the Economic Development Administration's grant programs due to redundancy with other federal economic development initiatives.
-

10.Census

- **Consolidation of Statistical Agencies: Merge the Census Bureau, Bureau of Economic Analysis, and the Bureau of Labor Statistics into a more streamlined, focused statistical agency.**

11.Bureau of Industry and Security (BIS)

- **Overruling Authority Change: Modify BIS's ability to overrule other agencies by making voting on licensing disputes equal among agencies, with unresolved cases elevated to high-level officials like the National Security Advisor.**
- **Data Reporting to Congress: Require BIS to provide quarterly data on export licenses to relevant congressional committees, enhancing oversight and accountability.**
- **Export Denial for Countries Lacking End-use Checks: Deny export licenses to countries like China and Russia that fail to provide adequate end-use verification.**
- **Export Control Modernization: Tighten U.S. export control policies for countries of concern, especially China and Russia, and reduce de minimize thresholds for critical technologies.**
- **Strengthening Entity List and Sanctions: Expand the Entity List for Chinese and Russian entities and impose a "policy of denial" for exports. Apps like TikTok and WeChat should be designated to the Entity List due to national security concerns.**
-

12.Minority Business Development Agency (MBDA)

- **Potential Elimination of Funding: There was a proposal to eliminate funding for the MBDA, which some view as fostering racial bias, with an eventual pivot to focus on policy rather than direct services.**

13. National Institute of Standards and Technology (NIST)

- **Privatization of Manufacturing Extension Partnership (MEP):** Fully privatize the MEP program that assists small and medium-sized manufacturers, removing government funding and transitioning operations to the private sector.
- **Transfer of Baldrige Performance Excellence Program:** Transfer the Baldrige Performance Excellence Program to the private sector, making it fully self-sustaining and managed by the Baldrige Award Foundation.

14. National Telecommunications and Information Administration (NTIA)

- **Big Tech Accountability for Censorship:** Propose policies to address censorship by big tech companies and defend free speech online.
- **Implementation of ICTS Executive Order:** Fully implement the ICTS Executive Order to eliminate national security threats in communication technology and services.
- **Expansion of Federal Spectrum Leasing:** Increase short-term leasing of government-allocated spectrum to optimize usage while protecting federal agency rights.
- **Support for Commercial Space Industry:** Advocate for licensing decisions that bolster U.S. leadership in the commercial space industry.
- **Review of FirstNet:** Evaluate the performance of FirstNet and its long-term viability in light of advancing technologies that may make it obsolete.

Treasury Department: Project 2025 Recommendations

Tax Reform

- **Repeal Inflation Reduction Act (IRA): This would increase costs for Americans and cut job creation. It also eliminates the tax credits and incentives associated with clean energy.**
- **Regressive Tax Code: Increase taxes for low- and middle-income Americans while cutting taxes for wealthy individuals.**
- **Corporate Tax Cuts: Reduce corporate taxes to 18% and reinstate conditions for offshore corporate tax havens.**
- **Eliminate Dual Mandate of the Federal Reserve: Remove the Fed's mission to maintain full employment.**
- **Dynastic Wealth: Reduce and cap estate taxes to benefit wealthier families.**
- **End SALT Deductions: Repeal state and local tax deductions, which would particularly impact blue states and cities with higher property taxes.**
- **End Clean Energy Tax Credits: Remove tax incentives for clean energy.**
- **Flat Consumption & Income Taxes: Support for a flat tax on consumption and income (excluding investments).**

IRS

- **Political Appointees: Fill the IRS with political appointees, avoiding Senate approval to reduce transparency and accountability.**
- **Reduce IRS Size: Propose cuts to the IRS.**
- **Eliminate Clean Energy Tax Incentives: Similar to the tax reform, push to eliminate clean energy incentives.**

OECD & International Affairs

- **End Support for the OECD: Cease U.S. financial support for the Organization for Economic Cooperation and Development.**

- **Withdraw from World Bank and IMF: Pull out of the World Bank and International Monetary Fund, ending financial contributions to both.**

Financial Regulation

- **Privatize Fannie Mae and Freddie Mac: Expose the housing market to predatory practices by privatizing these institutions.**
- **Repeal Corporate Transparency Act: Reverse efforts to improve transparency and fight money laundering by repealing the Corporate Transparency Act.**
- **Limit Employer Deductions: Reduce employers' ability to deduct the cost of worker benefits, such as health coverage for older dependents.**

Diversity, Equity, and Inclusion (DEI)

- **Eliminate DEI Offices: Abolish the Office for Diversity, Equity, Inclusion, and Accessibility, and the Counselor for Racial Equity within the Treasury.**
- **End DEI Practices: Eradicate any DEI-related practices at the Treasury Department.**

Overall Tax & Regulatory Policies:

- **Project 2025 seeks a regressive tax system that would disproportionately benefit the wealthy while imposing higher burdens on working-class Americans. It proposes reducing corporate taxes, eliminating incentives for clean energy, and repealing the Inflation Reduction Act, which would eliminate subsidies crucial for clean energy job creation and climate** action. Additionally, it calls for reducing the estate tax and offers easier paths for the wealthy to protect their assets.

- Consumption-Based Tax System: Shift toward a consumption-based tax system, with potential for national sales or cash flow taxes. This shift would likely increase the tax burden on low- and middle-income families.

Tax Code Revisions

- Two-Rate Individual Tax System: Implement a simple 15% and 30% individual tax structure, removing most deductions, credits, and exclusions.
- Corporate Tax Rate Reduction: Cut the corporate tax rate to 18% and apply a 15% rate for capital gains and qualified dividends.
- Immediate Expensing for Capital Expenditures: Allow immediate expensing for capital expenditures and adjust capital gains taxes for inflation.
- Repeal IRA Tax Increases: Repeal all tax increases enacted through the Inflation Reduction Act, including taxes on stock buybacks, coal, Superfund taxes, and taxes on drug manufacturers.
- Estate and Gift Tax: Lower the estate and gift tax to 20%, making permanent the 2017 tax bill's exemption increase for estates.
- Global Intangible Low-Taxed Income (GILTI): Reduce the tax rate on GILTI to 12.5% and eliminate the 20% haircut on related foreign tax credits.

SALT Tax Deduction

- Full Repeal: The temporary cap on state and local tax deductions should be fully repealed, affecting high-tax states like New York and California.

Education and Business Deductions

- **Repeal Educational Deductions: Eliminate tax deductions related to educational expenses.**
- **Business Preferences: Remove special business tax preferences, including deductions for energy-efficient commercial buildings.**

Universal Savings Accounts (USA)

- **Tax-Free Savings Accounts: Allow all taxpayers to contribute up to $15,000 of post-tax earnings into Universal Savings Accounts (USAs). These would function similarly to Roth IRAs, with non-taxable gains that could be withdrawn at any time for any purpose.**

Tax Legislation: Three-Fifths Vote Requirement

- **Supermajority Vote Requirement: Implement a rule that would require a 3/5ths majority in both the U.S. House and Senate to raise income or corporate tax rates.**

Global Minimum Tax Opposition

- **Opposition to Global Minimum Tax: Reject global minimum tax proposals, arguing they increase taxation and create a "global tax cartel."**

Consequences of Project 2025's Proposals:

- Personal Tax Regressivity: Slash personal income tax rates that benefit the wealthy, while eliminating key deductions for middle- and lower-income earners.
- Corporate Tax Cuts: Significant corporate tax cuts, particularly beneficial for large corporations, with minimal benefits for workers.
- Clean Energy Setback: The repeal of clean energy incentives would likely result in the loss of thousands of jobs, hinder competitiveness in global markets, and slow emissions reductions.
- Increased Wealth Inequality: The combination of tax cuts for the wealthy, the reduction in estate taxes, and the focus on offshore tax havens would result in more wealth concentration among the wealthiest Americans.
- Consumption Tax Shift: Moving to a consumption-based tax system would disproportionately harm low- and middle-income families, making it more difficult for them to afford basic needs.
- Reduced Regulatory Oversight: The repeal of the Corporate Transparency Act and deregulation efforts would make it easier for bad actors to engage in illicit financial activities like money laundering.

Businesses

- Business Loan Limitation & Net Operating Losses: Project 2025 advocates for an increase in the business loan limitation to at least $500,000 and suggests businesses should be allowed to fully carry forward net operating losses.
- Tax on Investment & Capital: The initiative proposes eliminating or reducing extra layers of taxes on investment and capital.
- Net Investment Income Surtax & Base Erosion Anti-Abuse Tax: Project 2025 calls for the elimination of both taxes.
- Cap on Untaxed Benefits: The next administration should set a cap (no higher than $12,000 annually) on untaxed benefits

that employers can claim as deductions. Employee benefits like gym facilities and other non-retirement expenses should be counted.

- Health Savings Accounts (HSA): Only a percentage of HSA contributions should count toward benefit deduction limitations, and these deductions should not increase with inflation.
- Employee Dependents: Employers should be denied deductions for health insurance and other benefits for employee dependents aged 23 or older.

IRS

- Form 1099-K Filing & $600 Threshold: Project 2025 opposes the requirement for banks to file a revised Form 1099-K for payments above $600, claiming it would create unnecessary administrative burdens.
- IRS Presidential Appointees: The project suggests increasing the number of presidential appointees at the IRS, including roles like Deputy Commissioners and the National Taxpayer Advocate, with some being exempt from Senate confirmation.
- IRS Budget & Office of the Taxpayer Advocate: It proposes maintaining the IRS operating budget constant (adjusted for inflation), increasing the Taxpayer Advocate's budget by 20%, and closing the Office of Equity, Diversity, and Inclusion (DEI).
- IT Modernization: Resources may be warranted for IRS IT modernization if effective plans and oversight are in place.

Organization for Economic Cooperation and Development (OECD)

- Project 2025 calls for the U.S. to withdraw from the OECD and stop funding it, criticizing it as a taxpayer-funded left-wing think tank that imposes regulations on low- or no-tax countries.

International Affairs

- Tax Assistance Protocol: Project 2025 recommends withdrawing the Protocol Amending the Convention on Mutual Administrative Assistance in Tax Matters, citing potential threats to privacy and security.
- Competent Authority Agreement: The next administration should oppose the Competent Authority Agreement on Automatic Exchange of Financial Account Information, asserting it could harm national security.
- World Bank & IMF: It proposes withdrawing from the World Bank and IMF, ending U.S. financial contributions, and calling for reforms to ensure these institutions align with American principles of free markets and limited government.

Committee on Foreign Investment in the U.S. (CFIUS)

- Project 2025 suggests that CFIUS establish a mitigation monitoring program, close the greenfield investments loophole, and develop a penalty schedule for violations related to foreign investments, particularly by Chinese-controlled entities.

Dodd-Frank Act

- Project 2025 advocates for the repeal of Title I, Title II, and Title VIII of the Dodd-Frank Act, which includes measures like the creation of the Financial Stability Oversight Council.

Fannie Mae and Freddie Mac

- It recommends moving toward the privatization of Fannie Mae and Freddie Mac to restore a sustainable housing finance market and reduce taxpayer reliance.

Financial Regulation

- Project 2025 supports merging regulatory bodies like the Office of the Comptroller of the Currency, FDIC, and National Credit Union Administration into a more streamlined entity.
- Corporate Transparency Act & FinCEN Rule: Calls for the repeal of the Corporate Transparency Act and the withdrawal of FinCEN's beneficial ownership reporting rule, which it argues would harm small businesses.

Diversity, Equity, Inclusion (DEI)

- Project 2025 calls for eliminating the Office of DEI, the Counselor for Racial Equity, and the Advisory Committee on Racial Equity within the Treasury Department. It opposes the practice of critical race theory and DEI, urging the next administration to reverse these initiatives.

Climate

- Project 2025 suggests eliminating the Treasury's Climate Hub Office and withdrawing from climate change agreements that it deems harmful to U.S. prosperity.
- It proposes promoting domestic energy investment, especially in oil and gas, and reversing efforts related to Environmental, Social, and Governance (ESG) principles that it claims have damaged U.S. energy security.

Trade

- Project 2025 calls for the elimination of the U.S. Trade and Development Agency, arguing that its functions should be handled by the private sector and that trade should be encouraged by reducing barriers and federal borrowing.

Financial Regulation

Federal Reserve

- Limit Independence & Autonomy: Proposes reducing the Federal Reserve's autonomy in managing key economic indicators and financial crises.
- Free Banking System: Suggests moving to a system where the Federal Reserve is abolished, replacing it with a decentralized banking system.
- Gold Standard: Advocates for a return to a gold standard, as was practiced in the 1800s (Suffolk System).
- Lender of Last Resort: Proposes limiting the Fed's ability to intervene in financial crises.
- Balance Sheet Wind-down: Suggests reducing the Federal Reserve's balance sheet, preventing its role in market stabilization.
- Dual Mandate: Calls for eliminating the Fed's dual mandate of both price stability and full employment, focusing on stable money instead.

SEC, CFTC, CFPB

- Weaken Public Company Reporting: Proposes reducing disclosure requirements for companies selling shares publicly, which could increase investment risks.
- Abolish CFPB: Seeks to dissolve the Consumer Financial Protection Bureau (CFPB), returning consumer protection duties to other agencies.
- Deregulate Cryptocurrency: Suggests easing regulations on cryptocurrency, exposing consumers to higher risks of fraud and loss.
- Reversal of Dodd-Frank: Proposes reversing parts of the Dodd-Frank Act, including consumer protections and financial stability measures.

Consequences of Project 2025's Financial Plans

- Weakened Market Stability: The Federal Reserve's reduced role could lead to a less stable economy, vulnerable to market risks and financial crises.
- Consumer Protection Rollback: Abolishing or weakening the CFPB would leave consumers at risk from exploitative financial practices.
- Cryptocurrency Risks: A lack of regulation could exacerbate fraud and illicit activity within the cryptocurrency space, risking consumer savings.

Housing and Urban Development

Key Proposals

- Reversing Fair Housing Regulations: Plans to undo Biden-era fair housing policies, including increasing mortgage insurance premiums and restricting eligibility for first-time homebuyer programs.

- Abolish PAVE Program: Seeks to eliminate the Property Appraisal and Valuation Equity (PAVE) program, which aims to address racial biases in home appraisals.
- Abolish Affirmatively Furthering Fair Housing: Proposes repealing the regulation ensuring federal funds are used to promote fair housing and combat segregation.
- HUD Overhaul: Suggests devolving many of HUD's functions to states, narrowing the department's scope, and limiting its role in housing and community development.

Personnel and Real Estate

- Political Appointments: Calls for increasing political appointments in leadership positions within HUD.
- Mortgage Insurance: Proposes increasing mortgage insurance premiums on products with terms over 20 years and maintaining premiums on refinances.
- Restricting Homebuyer Eligibility: Recommends narrowing eligibility for first-time homebuyers, with stricter qualifications.

Consequences of Project 2025's Housing Policies

- Increased Housing Costs: Higher mortgage insurance premiums would make housing less affordable for first-time buyers.
- Rollback on Housing Equity: The elimination of the PAVE program and Affirmatively Furthering Fair Housing would undermine efforts to reduce racial and ethnic disparities in housing.
- State-Level Control: Shifting HUD functions to states could lead to inconsistencies and unequal access to housing resources across regions.

Key Takeaways

- Financial Regulation: Project 2025's approach could reduce the Federal Reserve's ability to stabilize the economy, limit consumer protections, and deregulate financial markets, potentially leading to greater instability.
- Housing and Urban Development: The plan aims to shift away from federal oversight on fair housing, housing equity, and affordability programs, which could exacerbate inequality and housing challenges.

Climate Change

Key Proposals

- Repeal Climate Change Initiatives: Project 2025 calls for HUD leadership to repeal climate change-related initiatives and spending within the department's budget. This would remove any climate-focused policies or funding in HUD's operations and projects.

Consequences of Project 2025's Climate Policy

- Rollback of Climate Action: The elimination of climate change initiatives would reduce the focus on addressing environmental issues within housing and urban development, potentially undermining efforts to build sustainable, energy-efficient communities.
- Missed Opportunities for Climate Resilience: Without dedicated funding for climate-related initiatives, communities may lack the resources to adapt to

environmental risks like rising sea levels, extreme weather, or other climate change impacts.

PART XXII – Immigration

Key Proposals

- Prohibit Noncitizens from Federal Housing: Project 2025 proposes a regulation that would ban noncitizens, including mixed-status families, from living in federally assisted housing. This would reinstate a Trump-era policy and limit access to housing for immigrants and their families, regardless of legal status.
- Impact on Dreamers: The plan also suggests blocking financial aid for college students in states that offer in-state tuition to certain immigrant groups, including Dreamers (immigrants brought to the U.S. as children).
- End Legal Status for Dreamers: The proposal includes a move to terminate the legal status of around 500,000 Dreamers by halting the processing of renewal applications.
- Suspension of Temporary Worker Visa Updates: The administration would suspend updates to the eligible country lists for temporary worker visas (H-2A and H-2B), thus limiting the availability of foreign workers in key sectors like agriculture, construction, and hospitality.
- Bar Housing Subsidies for U.S. Citizens with Noncitizen Housemates: The plan proposes that U.S. citizens living with noncitizens or those without legal status would be disqualified from receiving federal housing assistance.
- Enforce State Cooperation on Immigration Data: States would be required to share information like driver's licenses and taxpayer ID numbers with federal authorities, or face risks to critical federal funding.

Consequences of Project 2025's Immigration Policy

- **Exclusion of Immigrant Families from Housing:** Banning noncitizens from federally funded housing could displace many immigrant families, particularly those in mixed-status households, reducing their access to safe, affordable housing.
- **Impact on Higher Education for Immigrants:** Blocking federal financial aid for students in states offering in-state tuition to immigrants could limit educational opportunities for many Dreamers and other legal immigrants.
- **Economic Impact on Critical Sectors:** Suspending updates to temporary worker visa lists and limiting immigrant access to key labor sectors could lead to labor shortages in agriculture, construction, and hospitality industries.
- **Increased Family Separation and Hardships:** Barriers to housing assistance for families with noncitizens could exacerbate economic hardship and family separations, creating instability for U.S. citizens living in mixed-status households.

Key Takeaways

These Project 2025 policies reflect a move towards prioritizing corporate and wealthy interests, which could have significant long-term consequences for middle-class economic stability and growth. There is nothing in the proposals for reducing the price of groceries or bringing down inflation (whereas tariffs would cause incredible inflation and a burden on consumers).

- **Climate Policy: The rollback of climate change initiatives within HUD would hinder efforts to create environmentally sustainable housing and urban infrastructure, potentially worsening long-term environmental and economic challenges.**
- **Immigration Policy: The proposed restrictions would have significant consequences for immigrant families and communities, excluding many from housing assistance, higher education, and vital labor markets, while increasing the role of state and federal control.**

PART XXIII - Transportation and Infrastructure

Renewable Energy and Infrastructure

Key Proposals

- **Halt IIJA and IRA Programs on Renewable Energy: Project 2025 recommends halting or stalling programs under the Inflation Reduction Act (IRA) and the Infrastructure Investment and Jobs Act (IIJA) that focus on renewable energy development. This would block the continued funding and expansion of clean energy initiatives, potentially slowing the transition to renewable energy sources.**
- **Rescind DOT's Grantmaking Ability: The proposal suggests stripping the Department of Transportation (DOT) of its grantmaking power and instead shifting this responsibility to Congress, allowing grants to be given directly to states.**
- **Abolish Discretionary Grantmaking at DOT: Project 2025 argues for the abolition of DOT's discretionary grant processes, suggesting that federal funding should focus solely on formula-based distributions to states. This would reduce**

federal involvement in how transportation funds are allocated and prioritize state-level decision-making.

Consequences of These Policies

- Reduced Federal Support for Renewable Energy: Halting or stalling renewable energy programs could significantly delay the adoption of clean energy infrastructure, potentially increasing reliance on fossil fuels and missing critical opportunities to combat climate change.
- State-Controlled Infrastructure Investments: Shifting grant-making power to Congress and eliminating discretionary grants would give states more control over infrastructure investments but could also lead to uneven development, with states prioritizing different projects based on political or economic preferences.

Emissions and Electric Vehicles (EVs)

Key Proposals

- Gut Emissions Regulations: Project 2025 calls for a rollback of Biden-era emissions standards, arguing that these regulations are ineffective at reducing global temperature trends.
- Reversal of EV Adoption Policies: The plan also advocates for reversing the current administration's pro-EV stance, claiming that increasing EV production makes the U.S. more dependent on foreign countries (particularly China) and strains the power grid.
- Lower Fuel Economy Standards: The proposal recommends lowering fuel economy standards to levels specified by Congress for 2020 vehicles (35 miles per gallon).

- Revocation of California's Waiver: Project 2025 advocates for revoking the special waiver granted to California, which allows the state to set stricter emissions standards than federal regulations.
-

Consequences of These Policies

- Stagnation of Clean Energy Efforts: Reversing emission regulations and halting the push for EV adoption could slow the transition to cleaner, more sustainable transportation, making it harder to achieve long-term environmental goals.
- Increased Dependence on Fossil Fuels: Rolling back fuel economy standards and emissions regulations could result in more pollution and higher reliance on oil and gas, potentially delaying the development of green energy alternatives.

Safety and Technology in Transportation

Key Proposals

- Remove Automated Vehicle Safety Measures: Project 2025 calls for removing certain safety measures related to the development and use of automated vehicles (AVs). It also proposes eliminating mandatory data collection for AVs and focusing instead on performance-based regulations.
- Support Joint Ventures in Aviation: The plan advocates for supporting joint-venture collaborations between smaller U.S. carriers like JetBlue and Spirit to help them scale and compete with larger carriers.
- FAA and ATO Reforms: Project 2025 proposes separating the Federal Aviation Administration (FAA) from the Department of Transportation (DOT) or at least separating the Air Traffic Organization (ATO) from the FAA. This would allow the FAA

to operate more like a business, with an emphasis on efficiency and cost control.

Consequences of These Policies

- Reduced Oversight for AV Safety: Less stringent safety regulations for automated vehicles could lead to more accidents or technological failures, potentially jeopardizing public safety as these technologies are developed.
- Weakened Oversight in Aviation: Separating the FAA from DOT could lead to a fragmented aviation oversight system, making it harder to ensure safe and coordinated operations across the aviation industry.
- Increased Competition in the Airline Industry: Supporting joint ventures between smaller carriers could increase competition in the airline industry, potentially lowering prices for consumers. However, this could also lead to reduced choices if consolidation reduces competition in specific markets.

Mass Transit and Rail

Key Proposals

- End Federal Support for Mass Transit Expansion: Project 2025 advocates for removing federal subsidies for mass transit expansion, focusing instead on maintaining and improving current facilities. It argues that transit expansion programs like the Capital Investment Grants (CIG) are wasteful and often uneconomical.

- Reform the Jones Act: The plan calls for reconsidering or repealing the Jones Act, which mandates that goods moved by water between U.S. ports be carried on U.S.-flagged ships. It argues that the Act creates inefficiencies and increases shipping costs.
- Railroad Crew Size Regulation: Project 2025 opposes the Federal Railroad Administration's proposed rule on crew size, claiming it reduces operational flexibility without improving safety.

Consequences of These Policies

- Decreased Investment in Public Transit: Cutting federal funding for mass transit expansion could hinder the development of new transit options, limiting access to affordable public transportation in some areas.
- Potential Shipping Inefficiencies: Repealing or significantly reforming the Jones Act could reduce shipping costs but might also undermine U.S. maritime jobs and industry, potentially weakening domestic shipping capacity.

<u>Overall Impact of Project 2025's Transportation Policies</u>

- Reduced Federal Investment in Infrastructure and Renewable Energy: Project 2025's recommendations would shift responsibility for transportation infrastructure to states and reduce federal funding for renewable energy and public transit programs. This could slow the country's transition to cleaner, more efficient transportation systems.
- Focus on State-Level Control: By emphasizing state-level decision-making, the plan could result in more tailored transportation solutions for individual states, but it could also

lead to disparities in transportation infrastructure quality and availability across the nation.

- Decreased Regulatory Oversight: The plan seeks to reduce federal regulatory oversight in areas like emissions standards and transportation safety, which could make it more challenging to address long-term environmental and safety concerns effectively.

PART XXIV - Elections

Campaign Finance and Federal Elections Commission (FEC) Reforms

Key Proposals

- Remove Contribution Limits for Donations: Project 2025 proposes removing or greatly raising contribution limits for donations, arguing that the current limits serve no practical purpose in preventing corruption and are an unnecessary constraint on candidates and political parties.
- Limit FEC's Enforcement Power: The plan calls for reducing the Federal Election Commission's (FEC) enforcement power, claiming that the agency's enforcement activities have been excessive and detrimental to small committees and campaigns. It argues that over-enforcement by the FEC has a "chilling effect" on political speech, especially for lower-level political activities.
- End USCYBERCOM's Election Protection Role: Project 2025 recommends ending U.S. Cyber Command's (USCYBERCOM) efforts to protect U.S. elections from foreign interference, claiming that such actions contribute to perceptions of partisan involvement by the Department of Defense.

Consequences of These Policies

Removal of Contribution Limits

- **Potential for Increased Political Influence: By eliminating or significantly raising donation limits, Project 2025 would make it easier for wealthy individuals and corporations to make larger donations to political candidates and parties. This could increase the influence of money in politics, potentially leading to a system where candidates are more beholden to their donors than to voters.**
- **Impact on Political Equity: Without contribution limits, the financial resources required to run competitive campaigns could disproportionately favor well-funded candidates, making it more difficult for candidates without substantial financial backing to compete.**

Weakened Oversight of Campaign Finance

- **Less Accountability: Reducing the FEC's power could lead to less accountability in political campaigns. While the FEC is tasked with enforcing laws that promote transparency and fairness in elections, weakening its enforcement could create opportunities for candidates and political committees to circumvent rules intended to prevent corruption or conflicts of interest.**
- **Risk of Corruption: If enforcement power is scaled back and contribution limits are lifted, it may be easier for special interests, wealthy individuals, and corporations to exert undue influence over the political process, potentially undermining the integrity of the electoral system.**

Ending Cybersecurity Measures for Elections

- **Increased Vulnerability to Foreign Interference: Ending USCYBERCOM's involvement in securing U.S. elections could leave the country more vulnerable to foreign interference, cyberattacks, and misinformation campaigns aimed at undermining the electoral process.**
- **Perceived Partisan Influence in Election Security: While the argument for ending USCYBERCOM's role is to eliminate the perception of partisan involvement, removing federal cybersecurity efforts could raise concerns about the government's commitment to safeguarding election integrity, especially in the face of increasingly sophisticated cyber threats from adversarial nations.**

Overall Impact of Project 2025's Campaign Finance and Election Security Proposals

- **Potential for Increased Money in Politics: Raising contribution limits and limiting enforcement power would likely tilt the political playing field further in favor of wealthy donors and large political organizations, potentially increasing the influence of money in elections and further entrenching existing power structures.**
- **Reduced Oversight and Transparency: By weakening the FEC's enforcement capabilities, the proposal could reduce transparency and accountability, making it harder to ensure that campaigns follow fair practices and don't violate election laws.**
- **Heightened Risk to Election Security: Ending USCYBERCOM's role in defending U.S. elections could leave the country's democratic process more vulnerable to manipulation or**

disruption by foreign adversaries, potentially diminishing public trust in election outcomes.

There is nothing in Project 2025 about guaranteeing lections will not be "rigged." Eliminating cybersecurity would do just the opposite of that. There is nothing about voter suppression, gerrymandering or anything else that could compromise election integrity.

PART XXV - Culture War

Reproductive Rights, Diversity, and Religious Policies in Project 2025

Key Proposals

- Extreme Abortion Bans: Project 2025 advocates for national extreme abortion bans, pushing for the most robust legal protections for the unborn that Congress can support. It calls for actions such as ensuring no employer is required to provide abortion accommodations or benefits, and using federal powers to restrict access to abortion, including preventing federal funding for states with pro-choice policies.
- Cutting Off Federal Funding from Pro-Choice States: The plan suggests using federal power to withhold funds from states that maintain pro-choice policies, including requiring states to report detailed abortion statistics or risk losing federal support.
- Eliminating Reproductive Health and DEI Terms from Federal Laws: It proposes removing terms like "diversity," "equity," "inclusion," "gender," and "reproductive rights" from federal

laws, arguing that such language infringes on First Amendment rights.

- Banning Reproductive Health Education in Schools: The proposal includes abolishing the teaching of reproductive health in schools, focusing instead on promoting pro-life views and family values.
- Banning Corporate and Federal DEI Programs: Project 2025 advocates for a nationwide ban on corporate and federal diversity, equity, and inclusion (DEI) programs, especially those that promote critical race theory (CRT) and gender ideology.
- Ban on Gender Confirmation Surgery for Minors: The plan calls for a ban on gender confirmation surgeries for minors, citing concerns about child welfare and labeling such procedures as forms of child abuse.
- Ban on Pornography and Legal Penalties for Producers: Project 2025 seeks to outlaw pornography, with legal consequences for its producers and distributors, including severe penalties for educators and public librarians involved in promoting it.
- Invoke the Insurrection Act to Quash Protests: The proposal includes invoking the Insurrection Act to suppress protests, signaling a strong crackdown on dissent and civil unrest.
- Infuse Christian Nationalism into Government: The project emphasizes the integration of Christian nationalist values into the highest levels of government, prioritizing faith-based initiatives and religious freedom in policy and international relations.

Consequences of These Policies

Impact on Abortion Rights

- Extreme Restrictions on Abortion Access: National extreme abortion bans would effectively limit access to abortion across the U.S., overriding state-level decisions on abortion access. This could lead to a reduction in reproductive autonomy for individuals, particularly in states that currently have protections for abortion access.
- Punitive Measures for Pro-Choice States: Cutting off federal funding to states with pro-choice policies could create economic and social tension, with vulnerable populations in those states potentially losing access to key federal programs and services.
- Restricted Health Benefits and Support: Requiring employers to provide equal benefits for pro-life support, while clarifying that no employer is required to provide abortion accommodations, could limit access to essential health care benefits for employees seeking reproductive care.

Diversity and Inclusion Measures

- Impact on Minority and Marginalized Communities: Banning DEI programs and removing terms such as "diversity" and "equity" from federal laws could undermine efforts to address systemic inequality and prevent future policies designed to support underrepresented communities. Without DEI initiatives, individuals from marginalized groups may face greater barriers to inclusion in education, employment, and other sectors.
- Erosion of Racial and Gender Justice Initiatives: The push to remove critical race theory (CRT) and gender ideology from curricula and federal policies could limit educational opportunities for students to learn about diversity, racial justice, and gender equality. This may hinder efforts to address ongoing social disparities and limit individuals' understanding of these critical issues.

Gender and LGBTQIA+ Rights

- **Criminalizing Gender Confirmation Surgery for Minors:** Banning gender confirmation surgery for minors would limit access to medical care for transgender youth, potentially leading to negative psychological and emotional consequences for individuals who feel their gender identity is at odds with their assigned sex at birth.
- **Increased Vulnerability for Transgender and Non-Binary Individuals:** The broader rhetoric around "child abuse" and restrictions on gender-affirming care could increase stigmatization and marginalization of transgender individuals, making it more difficult for them to access life-affirming medical and mental health services.

Pornography Ban

- **First Amendment Concerns:** Outlawing pornography and imposing severe penalties on its producers could have significant implications for free speech and personal liberties. The proposal also raises concerns about how it would be enforced, potentially criminalizing individuals who engage in or distribute adult content.
- **Potential for Overreach:** Targeting educators and public librarians as sex offenders for distributing materials related to sexual education or adult content could have chilling effects on freedom of expression, educational content, and access to information.

Protests and Civil Liberties

- **Crackdown on Dissent:** Invoking the Insurrection Act to shut down protests would likely curtail civil liberties and stifle dissent, potentially leading to widespread political repression and the suppression of free speech. This could create an environment of fear and limit public participation in democratic processes.

Christian Nationalism in Government

- Separation of Church and State: Integrating Christian nationalist values into government could blur the lines between church and state, potentially marginalizing non-Christian religious groups and individuals who do not adhere to these values. It could also restrict religious freedoms by prioritizing one religious worldview over others.
- International Relations: The prioritization of faith-based initiatives in U.S. foreign policy could alter relationships with non-Christian nations and potentially complicate diplomatic efforts aimed at global peace, development, and human rights.

Overall Impact

- Increased Control Over Personal Choices: These policies would grant the government increased control over individuals' personal choices, particularly in relation to reproductive health, gender identity, and sexual expression, potentially limiting freedoms and undermining privacy rights.
- Impact on Marginalized Communities: The potential removal of DEI programs, restrictions on abortion, and the criminalization of gender-affirming care could disproportionately harm marginalized groups, including women, LGBTQIA+ individuals, and people of color.
- Erosion of Democratic Freedoms: The proposal to use federal powers to suppress protests and dissent, combined with an emphasis on Christian nationalism, could undermine democratic freedoms, making it more difficult for citizens to exercise their First Amendment rights.

These proposals would significantly reshape American society, governance, and individual freedoms. And those who are criminalized for this or anything else might be deported to another country according an idea Trump keeps introducing along with the unconstitutional running for e third term.

PART XXVI – Bringing Down Inflation

Trump has inherited a strong economy from Biden in 2025, but people are suffering the decades' long effects of the American Dream being unattainable as prices increase, wages decrease, debt skyrockets and rich oligarchs have trillions in dollars while the rest of Americans remain (working) poor.

Trump promised he would bring down inflation. He actually said, "inflation will vanish completely."

Thus far, he has done the opposite or has promised to do the opposite of the known ways to bring down inflation.

Here's a list of strategies that can be used to reduce inflation:

1. Monetary Policy Tightening: Central banks can raise interest rates, making borrowing more expensive and reducing consumer spending and investment, which can help cool down inflation. Biden was slowly bringing this down after it was raised in the Pandemic years. Trump has promised to cut regulations, impose sweeping tariffs, overhaul energy policy and lower interest rates. He has also continually gone after DEI programs for why all these problems exist instead of focusing on the four year he was originally in office and mismanaged the Covid-19 Pandemic crisis. The U.S. went into a recession, which made the Fed increase interest rates to counteract the failing economy and the diminished power of the U.S. dollar.

2. Reduce Government Spending: Cutting government expenditures can decrease overall demand in the economy, which may help lower inflationary pressures. Trump has fired many federal employees, offered them a "buyout" that was never sanctioned (thus funded) by an act of Congress, and tried to unconstitutionally freeze funding on all federal grants and loans (already passed as law by Congress).

3. Increase Taxes: Raising taxes can reduce disposable income, which leads to decreased consumer demand and can help lower inflation. Trump has floated the idea of completely eliminating federal income taxes, which would bankrupt the entire country. Even if he does not eliminate income taxes, he and MAGA Republicans will cut taxes especially on the rich.

4. Increase Supply of Goods and Services: Encouraging the production of more goods and services, such as through technological advancements or reducing supply chain bottlenecks, can help meet demand and reduce inflation.

Trump wants to deregulate all businesses, but without nay focus on which ways this could counteract inflation. His promise to create a "Stargate" AI program seems at this point to be a cash grab to billionaires behind it. Even Elon Musk, deeply entrenched with Trump, has questioned its legitimacy.

5. Strengthening the Currency: A stronger currency can make imports cheaper, which reduces the cost of goods and services in the domestic market and can help reduce inflation. Tariffs, trade wars and outright invasions and conflicts with mostly allies like Canada, Mexico, Denmark, Colombia and NATO is Trump's only "negotiating tool," which has caused other countries to turn away from the United States as a trading partner.

6. Wage Controls: Limiting wage increases can prevent a wage-price spiral, where higher wages lead to higher production costs and, in turn, higher prices. But raising how much workers get can increase how much they can buy. That with limits on the price of goods could make people buy more. See next policy idea for more info.

7. Regulate Prices: Governments can implement temporary price controls on essential goods and services to prevent excessive price hikes. However, this can sometimes lead to shortages if not carefully managed. This would work if ally nations would partner with the United States to create a "firewall" against China and other nations who would try to create trade imbalances. This cannot be done with tariffs, especially against U.S. allies. Trump will not stop with his obsession with tariffs since he has said, "The word 'tariff' is the most beautiful word in the dictionary, even more beautiful than the word 'love'."

8. **Encourage Savings: Policies that encourage saving rather than spending can reduce demand in the economy, which can help control inflation. No one can save if everyone is being price-gouged by corporations and losing work (or being underemployed). Trump's policies will cause massive unemployment, increased prices on consumers (who actually pay the tariff for imported goods not the foreign countries importing them) and global instability (with allies turning away from American exported goods and services).**

9. **Improve Efficiency in Markets: Reducing inefficiencies in markets, such as monopolies or barriers to entry, can increase competition and reduce prices. Trump might actually believe his 1800s concept of tariffs and bringing countries' companies into the U.S. to produce goods, but that is archaic. In a 2000s global economy, many U.S. company moved their production outside America to pay less to workers or go to where foreign nations would benefit them more. It has to do with corporations, not just countries as in the olden days.**

10. **Strengthen Supply Chains: Improving supply chain logistics and reducing delays or disruptions can help lower costs, which can ultimately reduce inflation. This will not occur when Trump has not stopped the wars being waged by Russia and Israel as he promised he would on Day 1 of his administration. In fact, Trump has said he told the Chinese President Xi to negotiate with Russia's Putin to stop the war in the Ukraine. China is an ally nation to Russia and only benefits if the war continues and the United States fails.**

11. Control Speculation: Limiting speculative activity in markets, particularly for commodities like oil and food, can help prevent price bubbles and reduce inflationary pressures. Trump's continual and crazy executive orders have caused the stock market to plummet. This has been made worse by China releasing its AI called DeepSeek, which knocked down Big Tech stocks by 10% (or a trillion dollars in one day).

12. Promote Investment in Productive Capacity: Encouraging investment in sectors that increase the economy's productive capacity (such as infrastructure or technology) can help balance supply with demand over the long term. Trump ordered the Federal Reserve to lower interest rates, which does not lower inflation. It historically has caused more inflation. It is also unconstitutional for an U.S. president to order the Fed to do that.

PART XXVII – Trump's First 100 Days (2025)

Now, that Project 2025 is laid out for you, dear reader, let us look at a detailed analysis and prediction model. This is based on an AI (artificial intelligence) hypothesizing based on available data as of January 2025.

Day 1-10: Establishing Control & Executive Orders

- Day 1: Trump's return to office might begin with a large inauguration speech reiterating his "America First" mantra. He could immediately begin signing executive orders to undo

policies of the Biden administration, including reversing environmental regulations, reasserting energy policies focused on fossil fuels, and reinstating the "Muslim travel ban" or similar travel restrictions.

Trump has already started mass deportation of illegal migrants (that swept up legal citizens in its net), tried to unconstitutionally end birthright citizenship and freeze financial funding created by Congress, end DEI (diversity, equity and inclusivity) in government agencies, and pardoned all January 6th insurrectionists who tried to overthrow the U.S. government in 2021 regardless if they were violent criminals and hurt/ killed cops. There are over a hundred executive orders flooding the zone.

- **Day 2-5: Trump would likely focus on reorganizing his cabinet, appointing loyalists and allies to key positions. His administration could roll out new priorities, including economic recovery measures, perhaps pushing for tax cuts or tariffs aimed at "reviving" American manufacturing.**
- **Day 6-10: He might push through some form of immigration reform, aiming to restrict illegal immigration while calling for border wall construction. Trump's team could start laying the groundwork for a new healthcare proposal, attempting to repeal and replace the Affordable Care Act.**

Day 11-30: Messaging & Political Consolidation

- **Day 11-15: Trump's team may initiate media campaigns to highlight his administration's early victories, using rallies or social media to counteract negative press and energize his base. He would probably focus on shaping public perception and working closely with conservative outlets to build momentum.**
- **Day 16-20: Trump might initiate a public confrontation with "the swamp" (Washington insiders), reasserting his populist**

rhetoric. This could involve efforts to reduce regulations or challenge perceived bureaucratic inertia.

- Day 21-30: Major attention could be directed towards challenging what he sees as biased legal or election systems, including a push to reform or investigate election integrity, possibly through the establishment of new committees or even proposals for voter ID laws.

Day 31-60: Focus on Foreign Policy & National Security

- Day 31-40: Trump's foreign policy might shift back towards an "America First" stance, with a focus on renegotiating trade deals, especially with China, or pulling back from international commitments he views as disadvantageous to the U.S. He could prioritize military spending, positioning the U.S. to take a more aggressive stance on global security, particularly regarding Russia and China.
- Day 41-50: He might revisit the idea of withdrawing from multilateral agreements like the Paris Climate Accord, NATO, or other trade pacts. Trump could push for stronger alliances with countries that share his populist or nationalist views, like Hungary or Poland.
- Day 51-60: National security could also dominate his agenda, with possible military actions or interventions in response to threats. He may use his position to criticize global organizations like the UN, positioning the U.S. as a self-interested superpower.

Day 61-90: Domestic Policy and Legal Actions

- Day 61-70: Trump could announce a massive infrastructure plan, focusing on rebuilding roads, bridges, and energy infrastructure, with an emphasis on American jobs. This would likely come with a push for privatization of certain public assets, while simultaneously defunding social programs that are seen as inefficient.

- Day 71-80: Trump may take legal actions against members of the Biden administration or any opposition leaders, possibly using his powers to appoint special counsels or launch investigations into the conduct of his predecessors.
- Day 81-90: The Trump administration could focus on controversial social policies, perhaps pushing for a rollback of protections for LGBTQ+ communities or revisiting the role of religion in public life, like supporting religious liberty legislation.

<u>Day 91-100: Re-election Preparations & Culture War</u>

- Day 91-95: Trump would likely begin laying the groundwork for his 2028 re-election bid by reinforcing his base. He may hold rallies to promote his accomplishments and rally support, using these events to direct attention to ongoing culture wars. Expect criticism of "woke" culture, the media, and progressive movements.
- Day 96-100: By the end of the 100 days, Trump might signal a return to hardline positions on immigration, law and order, and economic nationalism. He could issue new policies or even executive orders in anticipation of a contentious re-election campaign in 2028.

Throughout these 100 days, it's likely Trump would balance significant executive actions with media-focused campaigns to energize his base, while focusing on challenging the political establishment and regaining his influence on the national and global stage. His agenda would probably focus heavily on national security, economic policies that favor his supporters, and creating a stark divide between "American interests" and globalism.

What then would the next four years look like under Trump's administration?

First let's look at the first Trump administration based on the facts and figures.

PART XXVIII – Trump's First Term (2017-2021)

As the curtain falls on Donald Trump's tenure as president, the data and statistics that define his time in office have been compiled, scrutinized, and finalized. These numbers, which serve as a historical record of the policies, decisions, and outcomes of his administration, paint a complex and multifaceted portrait of the past four years. Just as we did for his predecessor, we present a final look at the metrics that matter, offering a broad perspective on the state of the nation by the time Trump left office.

The Economy: A Mixed Bag

When it comes to the economy, the Trump years were marked by both successes and challenges. On the positive side, wages for American workers saw significant growth, outpacing inflation. Average weekly earnings increased by 8.4% after adjusting for inflation, a notable achievement during an administration that prioritized tax cuts and deregulation.

However, the economy experienced a major setback during Trump's presidency. The nation lost 2.7 million jobs, a sharp decline that was exacerbated by the COVID-19 pandemic. The unemployment rate rose by 1.7 percentage points to 6.4% by the end of his time in office, reflecting the toll that the global health crisis and subsequent economic disruptions had on the job market.

Despite these setbacks, some indicators pointed to an economy that was, at least in some areas, thriving. The stock market set new records, with the S&P 500 index soaring by 67.8%, a testament to the resiliency of the market and investor confidence during much of Trump's presidency.

On the flip side, the international trade deficit—an issue Trump campaigned on reducing—actually worsened during his time in office. By 2020, the U.S. trade deficit in goods and services had climbed to its highest level since 2008, increasing by 36.3% compared to 2016. This raised questions about the effectiveness of his trade policies, particularly his trade war with China, and their long-term impact on the economy.

Healthcare and Debt: Setbacks and Growth

Trump's promise to overhaul the U.S. healthcare system was never fully realized, and the results were apparent in the numbers. The number of people lacking health insurance increased by 3 million during his presidency, as efforts to repeal the Affordable Care Act (ACA) failed to materialize into substantial reform. While the administration pursued policies to limit the ACA's reach, including the removal of the individual mandate penalty, the broader promise of increasing affordable healthcare for Americans remained unfulfilled.

At the same time, the federal debt grew at an alarming rate. The national debt held by the public surged from $14.4 trillion in 2016 to $21.6 trillion by the end of Trump's presidency. This rise in debt was driven in large part by tax cuts, defense spending, and the economic fallout from the pandemic, which required extensive government spending on stimulus packages and relief measures.

Housing and Immigration: Shifts in Trends

The housing market experienced a remarkable boom under Trump's administration. Home prices rose by 27.5%, and the homeownership rate climbed by 2.1 percentage points to 65.8%, indicating that more Americans were able to purchase homes during this period. However, this surge in home prices also raised concerns about affordability, particularly for first-time homebuyers and those living in high-cost areas.

In terms of immigration, one of Trump's signature issues, the numbers were more contentious. While his administration made significant efforts to curtail illegal immigration, apprehensions at the Southwest border rose by 14.7% in

2020 compared to 2016. Despite his hardline policies, including the construction of a border wall and a crackdown on asylum seekers, the overall number of illegal immigrants apprehended at the border continued to climb, revealing the complexities of addressing illegal immigration.

Energy and Environment: Declines and Shifts

Trump's policies also had notable effects on the nation's energy landscape. Coal production saw a sharp decline of 26.5%, with coal-mining jobs dropping by 25%. While Trump had promised to revive the coal industry, this sector continued its long-term decline, driven by market forces and a shift toward cleaner energy sources.

However, on a positive note for the environment, carbon emissions from energy consumption dropped by 11.3%. This decrease came despite the administration's rollback of numerous environmental regulations and withdrawal from the Paris Agreement on climate change. The shift toward renewable energy sources and a broader national trend of decarbonization seemed to outweigh the administration's deregulatory push.

Gun Production and Crime: Unsettling Trends

In the realm of firearms, Trump oversaw a record-breaking year for handgun production, which rose by 12.5% in 2020 compared to 2016. This increase in gun production occurred amid growing concerns about gun violence and mass shootings, issues that remained highly contentious throughout his presidency. While Trump was a strong advocate for gun rights, the increase in firearms production coincided with a disturbing uptick in violent crime.

The murder rate, in particular, spiked to its highest level since 1997, raising alarm bells about the state of public safety in many cities across the country. Critics of Trump's handling of crime and policing pointed to the lack of substantial federal action to address rising crime rates, particularly in urban centers.

The Judiciary: A Lasting Legacy

One of the most enduring legacies of Trump's presidency was his impact on the judiciary. Trump successfully appointed one-third of the Supreme Court, reshaping its ideological makeup for years to come. He also appointed nearly 30% of the appellate court judges and a quarter of the district court judges, leaving a significant imprint on the federal judiciary. These appointments, many of whom were young and conservative, are likely to influence U.S. legal decisions for decades to come.

Trump’s Tariffs and Inflation: Economic Consequences and Warnings from Experts

During Donald Trump's presidency, the United States underwent significant shifts in its trade policy. One of the hallmark features of his administration was the imposition of tariffs, particularly on Chinese imports, as part of his broader "America First" strategy. While these tariffs were designed to strengthen American manufacturing, reduce trade deficits, and create jobs, they had unintended economic consequences, notably contributing to inflationary pressures across the economy. This chapter explores the effects of Trump's tariffs during both of his terms in office, the rationale behind them, and the warnings from economists about the potential inflationary fallout.

The Rise of Tariffs

Trump's tariff policy emerged from his belief that previous trade deals had been unfair to the United States. He argued that the U.S. was being exploited by trading partners like China, which had amassed a large trade surplus with the U.S., and that the trade balance needed to be corrected. As part of his "trade war," he imposed tariffs on a wide range of goods, especially those from China, in an attempt to force China to renegotiate trade terms and reduce its trade surplus with the U.S.

The most notable of these tariffs were those imposed in 2018, which targeted Chinese goods worth billions of dollars. Over the course of Trump's administration, the U.S. imposed tariffs on approximately $370 billion worth of Chinese goods, and China retaliated with tariffs on U.S. goods such as agricultural products. Trump's administration also imposed tariffs on steel and aluminum imports from several countries, including Canada, the European Union, and Mexico, citing national security concerns.

Impact on Inflation

While the goal of these tariffs was to encourage domestic production and protect U.S. workers, the immediate effect was an increase in prices for many goods. Tariffs are essentially a tax on imports, which means that foreign goods subject to tariffs became more expensive. American businesses relying on imported materials faced higher costs, which were often passed on to consumers in the form of higher prices for everyday products.

Inflation began to tick up during Trump's first term, and continued through his second, particularly after the tariffs were imposed. The increase in the cost of steel and aluminum, for example, led to higher costs for industries such as construction and automotive manufacturing. Similarly, tariffs on Chinese goods affected consumer electronics, furniture, clothing, and other products.

The inflationary effect of Trump's tariffs was compounded by several other factors, including supply chain disruptions, particularly as the COVID-19 pandemic exacerbated shortages of goods and raw materials. Nonetheless, economists widely agreed that tariffs contributed to inflationary pressure, especially on goods that relied heavily on imported inputs.

Economists' Warnings

Numerous economists, both domestically and internationally, warned against the use of tariffs, cautioning that the strategy could lead to higher prices and decreased economic efficiency. Leading figures such as Paul Krugman, a Nobel Prize-winning economist, were vocal critics of Trump's tariffs. Krugman argued that tariffs do little to solve underlying economic problems and instead raise the cost of living for consumers. In a 2018 New York Times column, Krugman described the tariffs as "a giant tax on consumers," noting that while they might provide short-term benefits to certain industries (like steel), they ultimately hurt the broader economy.

Other economists echoed similar sentiments. Former Federal Reserve Chair Janet Yellen expressed concerns that tariffs would stoke inflation, especially in a globalized economy where supply chains are intertwined. She warned that tariffs disrupt those supply chains and lead to inefficiencies, thus raising prices. In a 2019 testimony to Congress, Yellen explained that while tariffs might help some industries, they harm others, particularly consumers, who would ultimately pay higher prices for goods that had become more expensive due to trade barriers.

The Peterson Institute for International Economics, a respected think tank, also voiced strong opposition to the tariff strategy. The Institute argued that tariffs hurt American consumers by raising prices on essential goods, which disproportionately affected lower-income households. According to their studies, the tariffs imposed by the Trump administration would cost the average American household hundreds of dollars per year, while providing little tangible benefit in terms of job creation or wage growth in the long run.

Long-Term Consequences

The long-term consequences of Trump's tariff policy are still debated. While some industries, such as steel and aluminum production, saw a short-term boost, the broader economy experienced inflationary pressures. The U.S. dollar also experienced volatility as a result of the trade war, further contributing to instability in global markets.

One of the most profound effects of the tariffs was their impact on the global supply chain. Many U.S. companies, particularly those in manufacturing and technology, rely on components and raw materials from abroad. Tariffs disrupted these supply chains, making it more expensive for companies to produce goods domestically, and thus increasing prices for consumers. Economists warned that the long-term effect of this disruption would be a slowdown in global trade, reduced economic growth, and an overall less efficient global economy.

Conclusion: A Legacy Defined by Contradictions

Trump's tariffs were a bold attempt to reshape the U.S. economy by challenging long-standing trade practices and shifting the balance of trade. However, these tariffs also contributed to inflation, as they increased the cost of goods for consumers and disrupted established global supply chains. Many economists, including prominent figures like Paul Krugman, Janet Yellen, and those at the Peterson Institute for International Economics, warned against the inflationary risks of tariffs. While the goal of reducing trade deficits and revitalizing American manufacturing was understandable, the unintended consequence of rising prices for U.S. consumers underscored the complex trade-offs inherent in protectionist policies.

Ultimately, Trump's tariff policy serves as a cautionary tale: while tariffs can be an effective tool in addressing specific trade grievances, they also come with economic costs, especially when it comes to inflation and trade wars with both ally nations and foreign adversaries. The long-term effects of these policies will continue to be felt for years to come, as they have reshaped both domestic markets and international trade relationships.

In the final analysis, the statistics from Donald Trump's time in office reveal a presidency marked by both progress and setbacks, triumphs and failures. On the one hand, the economy saw growth in wages, stock markets, and homeownership, even as the trade deficit widened and the federal debt ballooned. On the other hand, the administration's handling of healthcare, immigration, crime, and environmental policy left much to be desired, with mixed results in these key areas.

Trump's impact on the judiciary will undoubtedly be one of his most lasting legacies, while his approach to governance—characterized by bold promises and contentious policies—

continues to provoke debate. As with all presidencies, history will be the ultimate judge of how these numbers and trends are interpreted, but for now, the data provides a comprehensive look at the state of America when Trump exited the White House.

Sources

1. Bureau of Labor Statistics. "Employment, Hours, and Earnings from the Current Employment Statistics survey (National); Total Nonfarm Employment, Seasonally Adjusted." Data extracted 6 Oct 2021.

2. 5 News. "Donald Trump: I will be the 'greatest president God ever created.'" 10 Feb 2016.

3. Robertson, Lori. "Trump Wildly Inflates Unemployment." FactCheck.org. 10 Feb 2016.

4. Reuters. "Janet Yellen Says the U.S. Is Close to Full Employment." Fortune. 18 Jan 2017.

5. White House. "Press Briefing by Press Secretary Sean Spicer." 10 Mar 2017.

6. Bureau of Labor Statistics. "Labor Force Statistics from the Current Population Survey; Unemployment Rate, Seasonally Adjusted." Data extracted 6 Oct 2021.

7. Bureau of Labor Statistics. "Job Openings and Labor Turnover Survey: Job Openings, Seasonally Adjusted." Data extracted 6 Oct 2021.

8. White House. "Economic Report of the President." Council of Economic Advisers. Feb 2020.

9. Jackson, Brooks. "Job Openings Bunk." FactCheck.org. 15 Nov 2018.

10. Farley, Robert. "Declining Labor Participation Rates." FactCheck.org. 11 Mar 2015.

11. Bureau of Labor Statistics. "Labor Force Statistics from the Current Population Survey; Labor Force Participation Rate." Data extracted 6 Oct 2021.

12. Bureau of Labor Statistics. "Labor Force Statistics from the Current Population Survey; Unemployment Level, Seasonally Adjusted." Data extracted 6 Oct 2021.

13. Bureau of Labor Statistics. "Labor Force Statistics from the Current Population Survey; All employees, thousands, manufacturing, seasonally adjusted." Data extracted 6 Oct 2021.

14. U.S. Bureau of Economic Analysis. "Table 1.1.1. Percent Change From Preceding Period in Real Gross Domestic Product." Interactive data extracted 6 Oct 2021.

15. Coibion, Olivier et al. "Labor Markets During the COVID-19 Crisis: A Preliminary View." National Bureau of Economic Research. Apr 2020.

16. Slotnick, David. "Boeing is temporarily closing its factories across Washington state after an employee died of COVID-19." Business Insider. 23 Mar 2020.

17. Krisher, Tom. "Automakers closing North American factories to protect workers from coronavirus." Associated Press. 18 Mar 2020.

18. Bureau of Labor Statistics. "Employment, Hours, and Earnings from the Current Employment Statistics survey (National); All employees, thousands, manufacturing, seasonally adjusted." Data extracted 6 Oct 2021.

19. The American Presidency Project. "Presidential Debate at the University of Nevada in Las Vegas." 19 Oct 2016.

20. Schwartz, Ian. "Trump: We're Going To See Economic Growth Of 4, 5 And Maybe 6 Percent." RealClear Politics 16 Dec 2017.

21. U.S. Bureau of Economic Analysis. Corporate Profits After Tax (without IVA and CCAdj) (CP), retrieved from FRED, Federal Reserve Bank of St. Louis. 6 Oct 2021.

22. Crane, Leland D. "Business Exit During the COVID-19 Pandemic: Non-Traditional Measures in Historical Context." Federal Reserve Board. Apr 2021.

23. McLaughlin, Patrick A., and Oliver Sherouse. RegData U.S. Regulation Tracker. QuantGov, Mercatus Center at George Mason University, Arlington, VA. Downloaded 5 Oct 2021.

24. McLaughlin, Patrick A. and Oliver Sherouse. RegData US 3.1 Annual (dataset). QuantGov, Mercatus Center at George Mason University, Arlington, VA. "RegData 3.1 Annual Summary." Downloaded 5 Oct 2021.

25. Federal Bureau of Investigation. "Crime in the United States 2020, Table 1." 27 Sep 2021.

26. U.S. Department of Agriculture, Food and Nutrition Service. "Supplemental Nutrition Assistance Program (Data as of Sept 3, 2021)." 6 Oct 2021.

27. U.S. Department of Agriculture, Food and Nutrition Service. Supplemental Nutrition Assistance Program (SNAP) data, FY 69 through FY21 National View Summary. ZIP Excel files. Accessed 6 Oct 2021.

28. Bureau of Economic Analysis. U.S. Trade in Goods and Services, 1960-present. Table 1. U.S. International Trade in Goods and Services. 2 Sep 2021.

29. Bureau of Economic Analysis. U.S. Trade in Goods and Services by Selected Countries and Areas, 1999-present. Table 3. U.S. International Trade by Selected Countries and Areas. 2 Sep 2021.

30. Kessler, Glenn. "Trump's nonsensical claim he can eliminate $19 trillion in debt in eight years." Washington Post. 2 Apr 2016.

31. Department of Treasury. Debt to the Penny. Debt Held by the Public. Fiscaldata.treasury.gov. Accessed 27 Sep 2021.

32. Department of Treasury. Final Monthly Treasury Statement. For Fiscal Year 2020 Through September 30, 2020, and Other Periods. Fiscal.treasury.gov. Oct 2020.

33.Office of Management and Budget. Historical Tables. Table 1.1—Summary of Receipts, Outlays, and Surpluses or Deficits (-): 1789–2026. Accessed 27 Sep 2021.

34.Office of Management and Budget. Historical Tables. Table 1.2—Summary of Receipts, Outlays, and Surpluses or Deficits (-) as Percentages of GDP: 1930–2026. Accessed 27 Sep 2021.

35.Office of Management and Budget. Historical Tables. Table 7.1—Federal Debt at the End of Year: 1940–2026 . Accessed 27 Sep 2021.

36.Congressional Budget Office. The Budget and Economic Outlook: 2020 to 2030. Cbo.gov. 28 Jan 2020.

37.Congressional Budget Office. The Budget and Economic Outlook: 2021 to 2031. Cbo.gov. 11 Feb 2021.

38.Energy Information Administration. U.S. Field Production of Crude Oil. 31 Aug 2021, accessed 29 Sep 2021.

39.Energy Information Administration. U.S. Imports of Crude Oil. 31 Aug 2021, accessed 29 Sep 2021.

40.Energy Information Administration. "Oil and petroleum products explained." 13 Apr 2021, accessed 29 Sep 2021.

41. Keisler-Starkey, Katherine and Lisa N. Bunch. "Health Insurance Coverage in the United States: 2020." Census Bureau. 14 Sep 2021.

42. Berchick, Edward R. and Heide M. Jackson. "Health Insurance Coverage in the 2017 CPS ASEC Research File." Census Bureau. undated.

43. Galewitz, Phil. "Breaking A 10-Year Streak, The Number Of Uninsured Americans Rises." Kaiser Family Foundation. 10 Sep 2019.

44. Centers for Disease Control and Prevention; National Health Interview Survey. "Health Insurance Coverage: Early Release of Estimates From the National Health Interview Survey, 2020." Aug 2021.

45. Centers for Disease Control and Prevention; National Health Interview Survey. "Health Insurance Coverage: Early Release of Estimates From the National Health Interview Survey, 2016." May 2017.

46. Lee, MJ et al. "GOP Obamacare repeal bill fails in dramatic late-night vote." CNN.com. 28 Jul 2017.

47.Kliff, Sarah. "Trump is slashing Obamacare's advertising budget by 90%." Vox.com. 31 Aug 2017.

48.Mukherjee, Sy. "The GOP Tax Bill Repeals Obamacare's Individual Mandate. Here's What That Means for You." Fortune. 20 Dec 2017.

49.Haberman, Maggie and Robert Pear. "Trump Sided With Mulvaney in Push to Nullify Health Law." The New York Times. 27 Mar 2019.

50.California v. Texas. No. 19–840. Supreme Court of the U.S. 17 Jun 2021.

51.Yahoo! Finance. "Dow Jones Industrial Average." Accessed 4 Oct 2021.

52.Yahoo! Finance. "S&P 500." Accessed 4 Oct 2021.

53.Yahoo! Finance. "NASDAQ Composite." Accessed 4 Oct 2021.

54.Egan, Matt. "The bull market turns 10 years old." CNN Business. 11 Mar 2019.

55.Menton, Jessica and Nathan Bomey. “Dow endures worst day since ‘Black Monday’; S&P 500 enters bear market as coronavirus spreads economic gloom.” USA Today. 12 Mar 2020.

56.U.S. Border Patrol. “Southwest Land Border Encounters.” Data accessed 7 Oct 2021.

57.U.S. Border Patrol. “Total Illegal Alien Apprehensions By Month Fiscal Years 2000-2019.” Undated. Accessed 7 Oct 2021.

58.Factba.se. “Remarks: Donald Trump Discusses the Border Wall in Alamo, Texas.” 12 Jan 2021.

59.Parks, Miles, Detrow, Scott and Snell, Kelsey. “Trump Signs Order To End Family Separations.” NPR. 20 Jun 2018.

60.McCullough, Jolie and Aguilar, Julian. “There are new, tougher guidelines for migrants seeking asylum in the United States.” The Texas Tribune. 12 Jul 2018.

61.Norwood, Candace. “Trump restricts immigration amid the pandemic. Critics see it as an excuse to push his own agenda.” PBS News Hour. 28 Jul 2020.

62. Farley, Robert. "Trump's Border Wall: Where Does It Stand?" FactCheck.org. 22 Dec 2020.

63. U.S. Customs and Border Protection. "Border Wall Status." 22 Jan 2021.

64. International Boundary and Water Commission. "Its Mission, Organization and Procedures for Solution of Boundary and Water Problems." Accessed 7 Oct 2021.

65. Rothbaum, Jonathan. "Census Bureau Still Studying Full Impact of Pandemic on Income Data." U.S. Census Bureau. 15 Sep 2020.

66. Shrider, Emily A., et al. "Income and Poverty in the United States: 2020." U.S. Census Bureau. 14 Sep 2021.

67. U.S. Census Bureau. "Table A-2, Households by Total Money Income, Race, and Hispanic Origin of Householder: 1967 to 2020." Sep 2021.

68. U.S. Census Bureau. "Table D-1. Historical Median Income Using Alternative Price Indices: 1967 to 2020." Sep 2021.

69. U.S. Census Bureau. "Median Household Income from 1967 to 2018 With and Without Adjusting for Differences from the

Income-Year 2013 Redesign and 2017 Processing System Update." Accessed 30 Sep 2021.

70. U.S. Census Bureau. "Figure 7. Percent Change in Median Earnings and Number of Workers: 2007 to 2009 and 2019 to 2020." Sep 2021.

71. U.S. Census Bureau. "Table B-4. Poverty Status of People by Family Relationship, Race, and Hispanic Origin: 1959 to 2020." Sep 2021.

72. Renwick, Trudi. "What is the Supplemental Poverty Measure and How Does it Differ from the Official Measure?" U.S. Census Bureau. 8 Nov 2012.

73. Renwick, Trudi and Liana Fox. "The Supplemental Poverty Measure: 2016." U.S. Census Bureau. 21 Sep 2017.

74. Fox, Liana and Kalee Burns. "The Supplemental Poverty Measure: 2020." U.S. Census Bureau. 14 Sep 2021.

75. U.S. Bureau of Alcohol, Tobacco, Firearms and Explosives. "Annual Firearms Manufacturing and Export Report, Year 2008 Final." 8 Mar 2011.

76. U.S. Bureau of Alcohol, Tobacco, Firearms and Explosives. “Annual Firearms Manufacturing and Export Report, Year 2016 Final.” 4 Jan 2018.

77. U.S. Bureau of Alcohol, Tobacco, Firearms and Explosives. “Annual Firearms Manufacturing and Export Report, Year 2019 Interim.” 7 Jul 2020.

78. U.S. Bureau of Alcohol, Tobacco, Firearms and Explosives. “Annual Firearms Manufacturing and Export Report, Year 2020 Interim.” 7 Jul 2021.

79. Brown, Dalvin. “Americans are loading up on guns and ammo in the wake of race protests.” USA Today. 3 June 2020.

80. Bartozzi, Joe. “TAKING STOCK OF RECORD-SETTING 2020 FIREARM YEAR.” NSSF. 7 Jan 2021.

81. Keegan, Desirée. “COVID-19 Spikes Gun Sales.” Dans Papers. 24 Mar 2020.

82. National Shooting Sports Foundation. “National Shooting Sports Foundation® Report: NSSF-Adjusted NICS – Historical Monthly Chart.” Proprietary data supplied on request and posted with NSSF permission. 5 Oct 2020.

83. "Gun shops in Los Angeles area attribute rise in sales to looting and protests." ABC 7. 2 Jun 2020

84. Bureau of Labor Statistics. "Employment, Hours, and Earnings from the Current Employment Statistics survey (National); All employees, thousands, coal mining," Data extracted 30 Sep 2021.

85. Bureau of Labor Statistics. "All Employees, Coal Mining." retrieved from FRED, Federal Reserve Bank of St. Louis. Updated 3 Sep 2021.

86. C-SPAN. "Presidential Candidate Donald Trump Rally in Phoenix, Arizona." 29 Oct 2016.

87. U.S. Energy Information Administration. "In 2020, U.S. coal production fell to its lowest level since 1965." 14 Jul 2021.

88. U.S. Energy Information Administration. "In 2020, the United States produced the least CO2 emissions from energy in nearly 40 years." 26 Jul 2021.

89. U.S. Energy Information Administration. Short-Term Energy Outlook Supplement: Forecast Sensitivity of Carbon Dioxide Emissions to Temperatures. Jul 2021.

90. U.S. Energy Information Administration. September 2021 Monthly Energy Review. 27 Sep 2021.

91. U.S. Energy Information Administration. "Table 11.1 Carbon Dioxide Emissions From Energy Consumption by Source." Sep 2021.

92. Bureau of Labor Statistics. "CPI for All Urban Consumers; All items in U.S. city average, all urban consumers." Data extracted on 30 Sep 2021.

93. Cox, Jeff. "Fed Chair Powell calls inflation 'frustrating' and sees it running into next year." CNBC. 29 Sep 2021.

94. Bureau of Labor Statistics. "The Economics Daily, Consumer Price Index rose 5.3 percent over the year ending August 2021." 17 Sep 2021.

95. Bureau of Labor Statistics. "CPI for All Urban Consumers; All items in U.S. city average, all urban consumers." Data extracted on 30 Sep 2021.

96. Bureau of Labor Statistics. "Employment, Hours, and Earnings from the Current Employment Statistics survey (National); Average Weekly Earnings of All Employees, 1982-1984 Dollars." Data extracted 1 Oct 2021.

97. Bureau of Labor Statistics. "Employment, Hours, and Earnings from the Current Employment Statistics survey (National); Average Weekly Earnings of production and nonsupervisory employees, 1982-1984 Dollars." Data extracted 1 Oct 2021.

98. Bureau of Labor Statistics. "Employment, Hours, and Earnings from the Current Employment Statistics survey (National); All employees, thousands." Data extracted 1 Oct 2021.

99. "US Business Cycle Expansions and Contractions." National Bureau of Economic Research. Updated on 19 Jul 2021.

100. University of Michigan Surveys of Consumers. "The Index of Consumer Sentiment." Accessed 1 Oct 2021.

101. University of Michigan Surveys of Consumers. "The Current Index of Consumer Sentiment." Accessed 1 Oct 2021.

102. University of Michigan Surveys of Consumers. "The Monthly Index of Consumer Sentiment." Accessed 1 Oct 2021.

103. Team Fix. "The CBS News Republican debate transcript, annotated." Washington Post. 13 Feb 2016.

104. Demsas, Jerusalem. "Covid-19 caused a recession. So why did the housing market boom?" Vox. 5 Feb 2021.

105. "Mortgage Rates Hit Record Low at Yearend." Press Release. Freddie Mac. 24 Dec 2020.

106. National Association of Realtors. "Sales Price of Existing Single-Family Homes." 22 Sep 2021.

107. National Association of Realtors. "Housing Statistic." Accessed 30 Sep 2021.

108. National Association of Realtors. "Methodology: Existing-Home Sales." Accessed 20 Sep 2021.

109. Rugaber, Christopher. "In a sizzling US market, demand and prices for homes soar." AP News. 25 May 2021.

110. U.S. Census Bureau. "Impact coronavirus." Accessed 20 Sep 2021.

111. U.S. Census Bureau. "Business and Industry." Accessed 1 Oct 2021.

112. Mateyka, Peter J and Christopher R. Mazur. "Homeownership in the United States: 2005 to 2019." U.S. Census Bureau. Mar 2021.

PART XXIX – The Next Four Years under Trump

Year 1: Reasserting Control & Policy Rollbacks

The first year would likely be a continuation of what we saw in the first 100 days, with an aggressive push to undo as much of the previous administration's policies as possible. We looked at Trump's first term to get a better data sample for the A.I. models. Here's a breakdown:

- Reversing Biden's Policies: Trump would focus heavily on reversing Biden-era legislation, particularly on climate change, healthcare, and immigration. This could include reimposing tariffs on China, dismantling parts of the Affordable Care Act, and pushing to finish or expand the border wall.
- Judicial Appointments: Trump would likely continue to focus on reshaping the judiciary, appointing conservative judges to lower courts. This would secure a lasting legacy, as those judges could remain in place for decades.
- Economic Nationalism: A renewed push for "America First" economic policies, including trade protectionism, tax cuts aimed at corporations, and potentially a reduction in social safety nets in favor of more defense spending and infrastructure projects.

- Foreign Policy Realignment: Trump would continue to scale back the U.S.'s involvement in global alliances and agreements, likely focusing on "America First" trade deals and a more isolationist foreign policy. Tensions with countries like China and Russia might escalate, while the U.S. pulls back from global interventions.

Year 2: Consolidation & Re-election Campaign Start

After his first year, Trump would likely start consolidating power and preparing for his re-election campaign. Note here, that the AI is unaware that Trump is not constitutionally allowed to run for another term since it would be his third. Trump has already begun talking about changing the U.S. Constitution, so he could run again. He may also try to claim a national emergency (like an "insurrection") to stay in power though it would be fabricated by him and his administration.

- Re-election Focus: Trump's second year would likely center on organizing his political base for the 2028 re-election bid. This could mean rallying his supporters, engaging in culture war issues (such as opposing progressive policies on gender, race, and social justice), and positioning himself as the champion of traditional American values.
- Law and Order: As part of his "tough on crime" stance, Trump might advocate for more police funding, anti-protest legislation, and a crackdown on what he sees as left-wing extremism. This could escalate the polarization between conservatives and progressives.
- More Executive Orders: Expect an increase in executive orders aimed at deregulation, particularly in areas like energy, environmental protections, and public education. This would likely include a focus on loosening restrictions on

fossil fuel industries and possibly scaling back environmental regulations.

- Conservative Social Policies: Trump's administration could focus on restricting abortion access, limiting transgender rights in schools, and pushing for more religious freedom legislation. These could be key issues in solidifying support from evangelical voters and conservative states.

<u>Year 3: Political Showdown & Economic Struggles</u>

Year three would likely be a turning point where Trump's administration faces mounting challenges, both politically and economically.

- Polarization and Pushback: As opposition to Trump grows, especially from the left and moderate Republicans, expect significant political gridlock. Trump would probably face pressure from both Congress and the courts on various issues, including attempts to limit or reverse the Affordable Care Act, and debates over immigration reform and his economic policies.
- Economic Pressure: Trump's push for tax cuts and deregulation might be met with an economic slowdown or backlash from critics who argue that his policies disproportionately benefit the rich. Trade wars could lead to price hikes or shortages in some industries. Public support may wane if the economy falters, and the GOP could experience internal divisions.
- International Tensions: Foreign policy could reach a breaking point as Trump's isolationist policies clash with global realities. Increased tensions with China, a potential escalation of conflicts in the Middle East, or unpredictable military confrontations could dominate headlines.

Year 4: Election Year Chaos & Legacy

In Trump's final year of his second term, the election cycle would kick into high gear. His focus would likely shift to securing his re-election, all while navigating the legacy of his presidency.

- Re-election Campaign: Trump would almost certainly begin campaigning for re-election in 2028 early, using rallies, social media, and his media allies to drive his message. Expect him to go full throttle on issues like "law and order," attacking "woke culture," and promoting the economy and job growth. Though he cannot legally run for reelection after serving two terms, there are Republicans trying to draft and pass legislation to change the Constitution Amendment to allow Trump to run for a third term.
- Scandals & Investigations: Depending on how the legal landscape unfolds, there may be ongoing investigations into Trump or his associates. These could focus on allegations related to the January 6th riot, financial dealings, or misuse of power during his first term.
- Increased Polarization: The rhetoric and political battles would intensify. Expect more heated clashes between the Trump administration and Democratic-controlled states or cities, as well as public debates over voting rights, election security, and civil rights.
- End of Term & Post-Presidency Plans: If Trump were to lose in 2028, he would likely make moves to consolidate his influence in the Republican Party and potentially influence the 2028 GOP primaries. If he wins, expect him to start another round of aggressive policy changes, potentially

cementing his position as the dominant figure in American politics.

Legacy and Long-term Impact

- Shaping the GOP: Trump's long-term influence on the Republican Party would likely be solidified, even if he doesn't run again. His populist, nationalist policies could have a lasting impact on future GOP candidates, and it's possible the party could become more focused on culture war issues, with less attention on traditional conservative policies like fiscal conservatism.
- Changing the Courts: His appointments to the federal courts could leave a significant conservative legacy, impacting decisions on issues like abortion, gun rights, and voting rights for decades.
- Divided America: His presidency would almost certainly continue to deepen the political divide in the U.S., particularly on issues like race, immigration, and climate change. The "Trump era" would be remembered as one of intense polarization and partisanship.

Overall, Trump's next four years could continue his push for populist, nationalist policies while simultaneously facing increasing opposition. His ability to maintain power would depend on his ability to keep his base energized and overcome the political and legal challenges that are likely to arise. As of this writing in 2025, Trump has said he does not care about the economy in an interview with Sean Hannity of Fox News. This is despite Trump promising to lower the price of groceries, bring down inflation and make living affordable for his base and all Americans (though he has repeatedly turned on those who he deems "the enemy within).

PART XXX – Trump's Malignant Narcissism, Sociopathy and Cognitive Decline

We shift from the general analysis and predictions of Artificial Intelligence to what the worst case scenarios might be. To do this, it requires understanding all the factors involved in the potential "algorithm" that factors in Trump's history, policy promises, adherence to Project 2025 and other ideas and ideologies. The A.I. began with Trump's family history and genetic predispositions affecting his decision-making and deviations potentially.

Trump has a family history of dementia, as well as, sociopathy including his father. He is almost 80 years old and is obese. He does not eat well, consuming fast food and doing only light exercise through golfing. He has a lifetime of actions that prove he is a sociopath with malignant narcissism, criminal behavior (that he was convicted of, including sexual abuse of woman and the 34 counts of falsifying business records to influence the 2016 U.S. Presidential Election results) and violent ideations (including the actual inciting of the riot that became the January 6th Insurrection on the U.S. Capital).

1. Speech and Communication Style:

- Trump has sometimes been noted for speaking in a manner that can seem fragmented, with his sentences occasionally lacking clear structure or logical flow. These instances may appear to some observers as signs of cognitive impairment, though others argue this could simply be his rhetorical style.

- He frequently repeats phrases and words during speeches, often relying on catchphrases like "tremendous," "great," and "fake news." While repetition isn't inherently a sign of cognitive decline, it can be a noticeable trait in his speech patterns.

2. Erratic Behavior or Memory Gaps:

- There have been moments during interviews or speeches where Trump seems to struggle with recalling specific details, such as names, facts, or events. For example, he has sometimes misstated dates or misremembered facts that are central to political discussions.
- In some cases, he has referred to events in ways that don't align with the general understanding of those events (e.g., making incorrect statements about past policies or outcomes), which critics sometimes point to as signs of memory issues or cognitive decline.

3. Health Assessments:

- During his first term as president, Trump underwent a public health assessment by his doctor, Dr. Ronny Jackson, who called Trump "mentally sharp." However, this evaluation was brief, and many critics have questioned the thoroughness of the evaluation, pointing out that it was not a comprehensive neurological exam.
- In the past, Trump has also made occasional references to his own health, often in a boastful way (for instance, discussing his fitness levels or comparing himself to others). Some people interpret this as deflecting questions about his health, though it doesn't offer concrete evidence of cognitive decline.

4. Public Scrutiny and Political Narrative:

- Given the contentious nature of Trump's presidency, claims of cognitive decline have often been used politically, particularly by his critics. For instance, during the 2020 election, Joe Biden's campaign and other critics raised questions about Trump's mental faculties, partly in response to his more controversial or unusual comments and actions.
- Some medical professionals who have been critical of Trump have pointed to aspects of his behavior, such as difficulty staying on topic or engaging in detailed policy discussions, as potential evidence of cognitive decline. However, such assessments have not been backed by formal testing or public diagnoses.

5. Age and Cognitive Decline in General:

- Trump was born in 1946, which places him in his late 70s as of the 2024 election. Cognitive decline naturally becomes a concern for many people as they age. However, cognitive decline is a broad and complex issue, and it doesn't necessarily manifest in obvious ways in all individuals.

Ultimately, whether or not Donald Trump is experiencing cognitive decline is a subject of speculation and debate. Cognitive issues can be complex and subtle, and without a formal, detailed evaluation by a medical professional, it's impossible to make definitive claims.

Sociopathy and Cognitive Decline in a Leadership Context

Sociopathy (or antisocial personality disorder) is typically characterized by a lack of empathy, disregard for the rights of others, and impulsive or manipulative behavior. If someone in a leadership position exhibits sociopathic traits, the effects on governance could be significant, especially if those traits are combined with cognitive decline. Let's explore how these factors could affect Trump during a second term.

1. Impulsivity and Decision-Making

- Sociopathy: Trump's history of impulsive decision-making, disregard for established norms, and sometimes erratic behavior could become more pronounced if his sociopathic traits were intensified. Sociopaths often make decisions based on personal gain or immediate gratification rather than long-term considerations. This could lead to erratic and unpredictable policies, with little regard for the broader impact on society.
- Cognitive Decline: Cognitive decline, particularly issues with memory, attention, and processing speed, could exacerbate this. Cognitive decline could result in impaired judgment or slower responses to emerging issues. This would likely affect his ability to engage in complex diplomatic negotiations, handle crises, or even make informed decisions on domestic policies. His inability to answer questions with simple sentences, to complete a sentence without going on tangents and diatribes and his non sequiturs via tweets and interviews gives basis to this decline.

2. Relationships with Allies and Diplomacy

- Sociopathy: A lack of empathy and disregard for the feelings or needs of others could strain relationships with allies. Trump's transactional approach to foreign policy often involved making deals based on what he believed would benefit him or the U.S. in the short term, without always considering long-term diplomatic relationships or the wellbeing of foreign populations. If his sociopathy were more pronounced, he might become even more isolative, pushing countries to align with him under threat rather than mutual benefit.

- Cognitive Decline: Cognitive decline could impair his ability to understand the nuances of international relationships or engage in high-level diplomatic talks. It may become more difficult for him to retain complex information about other countries' political climates, histories, or international agreements, leading to strained or even deteriorated relationships with allies and adversaries.

3. Public Communication and Trust

- Sociopathy: Sociopathic tendencies like lying, manipulation, and a lack of concern for the truth can erode public trust. Trump has been known to frequently make exaggerated or false statements, and his inability to apologize for mistakes has contributed to a divisive public image. In a second term, his sociopathy could lead to even more disregard for truth in favor of self-preservation and political expediency. This could increase public polarization and damage the credibility of his administration.
- Cognitive Decline: Cognitive decline could further complicate public communication. If Trump's cognitive abilities diminished, he might struggle with maintaining coherence during speeches or interviews. This could increase the perception of instability and untrustworthiness, especially if his public speaking became more disjointed or if he had difficulty keeping track of policy positions or promises.

4. Management of Internal Conflicts

- Sociopathy: Sociopaths tend to have an "us vs. them" mentality and often escalate conflicts. In a second term,

Trump might intensify political polarization by framing the opposition as enemies to be defeated rather than political rivals to be negotiated with. He might also surround himself with loyalists and make more impulsive personnel decisions, firing or demoting people who disagree with him. This could create a lack of diverse perspectives within the administration and lead to dysfunction in government.

- Cognitive Decline: If cognitive decline set in, Trump's ability to manage internal conflicts could be severely impaired. Difficulty processing complex information or understanding differing viewpoints could result in poor decision-making, leading to internal chaos. His reaction to challenges from Congress, the judiciary, or his own cabinet might become more erratic or unpredictable.

5. Public Perception and Mental Health

- Sociopathy: If Trump's sociopathic traits become more apparent, public perception of his character could worsen. His lack of empathy, focus on self-interest, and potential manipulative behavior might alienate moderates and swing voters, but could strengthen his appeal to his core base, which tends to appreciate his unapologetic, combative style.
- Cognitive Decline: Cognitive decline might be harder for the public to ignore, especially as age-related cognitive issues typically become more apparent. If his mental decline became visible (through lapses in memory, incoherent speech, or erratic behavior), it could prompt calls for him to step down or raise concerns about his fitness for office. Even those within his base who remain loyal might begin to question whether he is capable of fulfilling the demands of the presidency.

6. Legal and Ethical Concerns

- **Sociopathy: Sociopaths often disregard laws and ethical guidelines if they conflict with their desires or goals. If Trump's sociopathy manifests in more extreme ways, it could lead to more aggressive, controversial actions that might skirt legal boundaries. For example, he might continue to challenge norms or laws that inhibit his ability to act as he wishes, possibly leading to more clashes with Congress or the judiciary.**
- **Cognitive Decline: As cognitive decline progresses, Trump might lose the ability to effectively navigate complex legal issues. His understanding of legal constraints on presidential power might become muddled, leading to potentially reckless actions. If he were involved in any legal battles (such as those related to investigations from his first term), cognitive decline could complicate his defense or ability to engage effectively in his legal strategy.**

Conclusion

If Trump were to experience a combination of sociopathy and cognitive decline during a second term in office, his leadership could become even more erratic, unpredictable, and potentially harmful both domestically and internationally. His sociopathic tendencies might lead to more aggressive, self-serving policies with little regard for the long-term consequences, while cognitive decline could impair his ability to process complex issues, engage in meaningful diplomatic efforts, and communicate effectively with the public. The result could be a presidency marked by increased dysfunction, polarization, and instability. His base might remain loyal, but broader support could erode, potentially undermining his ability to lead effectively.

Lest we forget, 70,000+ health professionals signed a 2017 petition regarding Trump's mental stability and called for him to be removed from the Oval Office during his first term as U.S. President:

"We, the undersigned mental health professionals (please state your degree), believe in our professional judgment that Donald Trump manifests a serious mental illness that renders him psychologically incapable of competently discharging the duties of President of the United States. And we respectfully request he be removed from office, according to article 4 of the 25th amendment to the Constitution, which states that the president will be replaced if he is "unable to discharge the powers and duties of his office."

PART XXXI – Trump's 2016 Election and Russia's Interference

The Mueller Report, officially titled *Report on the Investigation into Russian Interference in the 2016 Presidential Election*, was a detailed investigation led by Special Counsel Robert Mueller into Russian interference in the 2016 U.S. presidential election and potential links between the Trump campaign and Russian officials. The report, released in two volumes in 2019, provides a comprehensive overview of the findings, detailing both Russian election interference efforts and possible obstruction of justice by then-President Donald Trump.

Volume I: Russian Interference in the 2016 Election

The first volume of the report examines Russian interference in the 2016 election, focusing on two main areas:

1. Russian Social Media Campaign: The Russian government, specifically the Internet Research Agency (IRA), engaged in a broad campaign to influence U.S. politics through social media. This included creating fake social media accounts to promote divisive content, targeting particular demographic groups, and spreading misinformation with the goal of undermining Hillary Clinton's campaign and supporting Trump.
2. Russian Hacking and Dissemination of Information: The Russian military intelligence agency, GRU, hacked into Democratic National Committee (DNC) servers and the Clinton campaign's emails, releasing stolen information through platforms like Wikileaks to damage Clinton's candidacy. The report finds that while the Russian interference was extensive, there was insufficient evidence to establish that members of the Trump campaign conspired or coordinated with the Russian government in these efforts.

However, the report also documents multiple contacts between Trump campaign officials and individuals with links to the Russian government. These interactions raised concerns but did not meet the legal threshold for conspiracy or criminal coordination. The investigation did not find conclusive evidence that the Trump campaign "colluded" with Russia, a central question of the probe.

<u>Volume II: Obstruction of Justice</u>

The second volume of the report focuses on whether President Trump obstructed justice during the investigation. Mueller's team examined several episodes involving Trump's actions, including his attempts to influence the investigation, his firing of FBI Director James Comey, and his efforts to limit or end the Russia investigation.

The key findings include:

1. Trump's Efforts to End the Investigation: The report highlights several instances where Trump attempted to end or limit the investigation. One of the most notable incidents involved Trump trying to have special counsel Mueller removed from the investigation and later pushing for the investigation to be concluded quickly, with several communications and directives indicating a desire to halt the probe.
2. Firing of James Comey: The report also details Trump's firing of FBI Director James Comey, which occurred after Comey refused to publicly state that Trump was not under investigation. While the firing was seen as an attempt to influence the investigation, Mueller's team did not find conclusive evidence to charge Trump with obstruction.
3. Inability to Conclude on Obstruction: Ultimately, the report concluded that there was insufficient evidence to charge Trump with obstruction of justice. However, Mueller emphasized that the investigation did not exonerate the president either, and legal scholars could reasonably interpret the findings in different ways. Mueller's team left the decision on obstruction to Congress, citing DOJ guidelines that prohibit indicting a sitting president.

Conclusion and Aftermath

The Mueller Report did not establish that the Trump campaign coordinated with Russia to influence the election, but it confirmed significant Russian interference. The report also highlighted numerous instances where Trump attempted to obstruct the investigation, though it ultimately declined to reach a conclusion on whether those actions amounted to a crime.

The release of the report led to debates over the implications of the findings, with Trump and his supporters claiming it as a vindication, while critics pointed to the report's detailed accounts of questionable actions. Mueller himself testified before Congress in July 2019, further clarifying the report's findings and emphasizing that obstruction remained an open question.

The report has been an important document in the ongoing scrutiny of the Trump presidency, influencing public opinion, legal analyses, and congressional investigations. It is also impossible to know how much Trump, the Barr-led DOJ and others in the administration influenced the very report of Mueller since he worked for them.

PART XXXII - Chapter: Trump's Impeachment Trials

The impeachment trials of Donald J. Trump represent a remarkable chapter in American political history, marked by unprecedented events and controversies that divided the nation. Trump became the first president in U.S. history to be impeached twice, each trial carrying with it its own unique circumstances and political tensions. The two impeachments—one in 2019 and the second in 2021—were tests not only of Trump's leadership but also of the nation's political and constitutional fabric. This chapter delves into the origins, proceedings, and outcomes of these historic trials.

The First Impeachment (2019)

The Allegations

The first impeachment of Donald Trump was initiated in December 2019, driven largely by the findings of a whistleblower complaint and an investigation by the U.S. House of Representatives. The core issue stemmed from a phone call Trump had with Ukrainian President Volodymyr Zelensky on July 25, 2019. In that conversation, Trump was accused of withholding nearly $400 million in military aid to Ukraine as leverage to press Zelensky into launching an investigation into Joe Biden, who was running for president at the time, and his son Hunter Biden's business dealings in Ukraine.

The House of Representatives, led by Speaker Nancy Pelosi, conducted an inquiry into these actions, with a particular focus on whether Trump's behavior constituted abuse of power. The allegations also included obstruction of Congress, as Trump directed his administration to defy subpoenas and withhold testimony and documents from the impeachment inquiry.

The Articles of Impeachment

On December 18, 2019, the House of Representatives voted to impeach Trump on two charges: abuse of power and obstruction of Congress. The vote saw near-unanimous support from Democrats and unanimous opposition from Republicans. Trump became the third president in U.S. history to be impeached, after Andrew Johnson (1868) and Bill Clinton (1998).

The Senate Trial

The impeachment then moved to the Senate for a trial in early 2020. The Senate, controlled by Republicans, was widely expected to acquit Trump, as no Republican senators seemed inclined to vote for his removal from office. The trial was marked by partisan gridlock, with Republicans calling for a quick dismissal and Democrats pushing for additional witnesses and documents.

The trial proceedings began in January 2020, and Trump's defense team argued that the charges against him were politically motivated and did not meet the constitutional standard for impeachment. They further contended that Trump's request for Ukraine to investigate the Bidens did not amount to an abuse of power, nor was it an impeachable offense. On the other hand, House Democrats made their case that Trump's actions undermined national security and the integrity of the 2020 election.

One of the key moments of the trial came when Senate Majority Leader Mitch McConnell announced that Republicans would not allow witnesses, a decision that enraged Democrats and many political commentators who felt it deprived the trial of credibility and transparency. The Republicans held firm, and after a two-week trial, the Senate voted to acquit Trump on both charges on February 5, 2020. The final vote was largely along party lines, with Mitt Romney, the Republican senator from Utah, casting the only vote to convict Trump on the abuse of power charge.

Aftermath of the First Impeachment

The impeachment and subsequent acquittal did little to damage Trump's political standing with his core base. In fact, many of his supporters viewed the impeachment process as a partisan witch hunt. However, the trial did have an impact on the broader political landscape, contributing to the intensifying polarization between the two major parties. It also set the stage for the 2020 election, in which Trump's handling of impeachment became a major talking point.

The Second Impeachment (2021)

The Events Leading to the Impeachment

The second impeachment of Donald Trump came in the wake of the January 6, 2021, Capitol riot. On that day, a mob of Trump supporters stormed the U.S. Capitol building in an attempt to overturn the results of the 2020 presidential election, in which Joe Biden defeated Trump. The riot followed weeks of Trump's false claims that the election had been "stolen" from him.

On January 6, 2021, Trump addressed a rally of his supporters, repeating the false narrative of a stolen election and encouraging them to "peacefully and patriotically make your voices heard" at the Capitol. The ensuing chaos led to the deaths of five people, including a Capitol police officer, and significant damage to the Capitol building. In the aftermath, Trump was accused of inciting an insurrection against the government.

The House of Representatives acted swiftly, drafting a single article of impeachment against Trump for "incitement of insurrection." On January 13, 2021, just a week before Trump's term ended, the House voted 232-197 to impeach him again, making Trump the first president in U.S. history to be impeached twice.

The Senate Trial

The Senate trial for Trump's second impeachment began on February 9, 2021, after he had already left office. A key legal question was whether a former president could be tried for impeachment, and the Senate ultimately ruled that the trial could proceed, with 56 senators voting in favor of the trial's legitimacy.

Trump's defense team argued that he was protected by the First Amendment and that his words did not incite the violent events that followed. They also argued that impeachment was unconstitutional because Trump was no longer president. On the other hand, House Democrats contended that Trump's rhetoric and actions directly incited the mob to storm the Capitol and disrupt the certification of the election results.

After days of emotional testimony, including video footage of the riot and accounts from lawmakers who had been in the Capitol during the attack, the Senate voted on February 13, 2021. The vote was 57 guilty to 43 not guilty, falling short of the required two-thirds majority (67 votes) to convict Trump. Seven Republican senators joined Democrats in voting to convict, making it the most bipartisan impeachment vote in history. However, Trump was acquitted for the second time.

Aftermath of the Second Impeachment

While Trump's second impeachment trial ended in acquittal, it was significant in several ways. It demonstrated a fracture within the Republican Party, as a handful of GOP senators voted to convict Trump for his role in inciting the insurrection. Moreover, it underscored the deep divisions in the country, as the events of January 6 left a lasting scar on the nation's political psyche.

Though Trump was acquitted, the trial had long-term consequences for the Republican Party. It led to a public reckoning within the party, with some distancing themselves from Trump, while others continued to embrace his influence. The impeachment also set the stage for Trump's post-presidency, as he remained a dominant figure in American politics, often threatening to run for president again in 2024.

Conclusion

The impeachment trials of Donald Trump were landmark events that forever altered the landscape of American politics. The first trial, centered on his dealings with Ukraine, reflected broader concerns over presidential abuse of power. The second trial, following the Capitol insurrection, was a stark reminder of the dangers of political extremism and the fragility of democratic institutions.

Both trials showcased the power of impeachment as a constitutional tool, but also revealed the deep partisan divides in Congress and the country. Trump's acquittals, despite overwhelming public attention and widespread condemnation of his actions, highlighted the limits of impeachment as a check on executive power. In the end, Trump's impeachment trials served as a defining moment in his presidency, in American political history, and in the ongoing debate over the proper role of impeachment in the U.S. system of government.

PART XXXIII – Trump's 34 Felony Convictions

President Donald Trump became the first ex-president in U.S. history to be convicted of felony crimes. A New York jury found him guilty of all 34 charges related to a scheme to influence the 2016 election by making a hush money payment to porn actress Stormy Daniels, who claimed they had an affair. Trump remained stoic as the verdict was read, and his legal team immediately signaled an intention to appeal the decision.

The case revolved around Trump's former lawyer and "fixer," Michael Cohen, who had facilitated the hush money payment and later cooperated with prosecutors. Cohen's testimony was critical in linking Trump directly to the scheme, revealing how the payments were made to silence Daniels and other women with alleged stories of affairs with Trump, just before the election. The defense countered by claiming the payments were legitimate legal expenses, and that any hush money deals were motivated by personal concerns, not political ones.

Despite the conviction, the political fallout remains uncertain though Trump was allowed to run for office and retake the seat of power. He was never pardoned though he believes he can pardon himself since the Supreme Court has upheld presidential immunity. While Trump's allies, including many Republican leaders, maintained their support, the case added further drama to his already controversial political career. His legal troubles are far from over, with other felony cases still pending, including those related to his efforts to overturn the 2020 election and mishandling classified documents.

Trump responded to the verdict by calling the trial "rigged" and emphasizing that the final decision would be made by voters in the upcoming November election. The trial's outcome has raised questions about whether a conviction could influence Trump's presidential bid, but his campaign has already framed the conviction as a politically motivated attack. As Trump faces sentencing in July, the timing of this legal drama could have a significant impact on his 2024 election campaign, even if it doesn't necessarily change his base of support.

In sum, this chapter marks a major milestone in Trump's legal battles, with the potential for long-term political implications. However, Trump's defiance and the backing he continues to receive from many Republicans suggest that the conviction may not substantially alter the dynamics of the 2024 election, where issues like inflation and broader policy concerns are likely to dominate.

PART XXXIV – Trump's Controversial and Criminal Acts and Ideas

Donald Trump has been involved in numerous controversial statements and actions throughout his career, particularly during his presidency and beyond. Here's a list of some of the most notable controversies:

1. "Access Hollywood" Tape (2005)

- In 2016, a recording surfaced where Trump was heard making lewd comments about women, saying, "Grab them by the pussy," while talking to Billy Bush. He dismissed it as "locker room talk," but the comments were widely criticized.

2. Impeachment (2019)

- Trump was impeached twice by the House of Representatives. The first impeachment was over abuse of power and obstruction of Congress related to his dealings with Ukraine. The second impeachment followed the January 6th Capitol riot, with charges of incitement of insurrection. He was acquitted in both Senate trials.

3. Charlottesville Response (2017)

- After a white nationalist rally in Charlottesville, Virginia, turned violent, Trump said there were "very fine people on both sides," drawing criticism for appearing to equate white supremacists with anti-racist protesters.

4. Muslim Travel Ban (2017)

- Early in his presidency, Trump signed an executive order banning travel from several predominantly Muslim countries, which critics viewed as discriminatory and Islamophobic.

5. Referring to African Countries as “Shithole” Countries (2018)

- Trump allegedly referred to Haiti, El Salvador, and African nations as “shithole countries” during a meeting on immigration. His remarks were condemned as racist.

6. Handling of COVID-19 (2020)

- **Trump was criticized for downplaying the severity of COVID-19 early in the pandemic, suggesting that the virus would "disappear," and for inconsistent messaging regarding mask-wearing and social distancing.**

7. Attacks on the Media

- **Trump repeatedly referred to the press as the "enemy of the people" and frequently called news outlets like CNN and The New York Times "fake news" whenever they published stories critical of him.**

8. Separation of Migrant Families (2018)

- **His administration implemented a "zero tolerance" policy that resulted in the separation of thousands of children from their parents at the U.S.-Mexico border. The policy was widely condemned as inhumane.**

9. Donald Trump's Remarks on the COVID-19 Vaccine (2021)

- **Although he had promoted the development of the COVID-19 vaccine under Operation Warp Speed, Trump also made controversial statements regarding vaccine efficacy and safety, sometimes sowing doubt about vaccines in the public's mind.**

10. The Insurrection and January 6th Capitol Riot (2021)

- Trump's rhetoric and claims about election fraud played a significant role in inciting the January 6th storming of the U.S. Capitol by his supporters. His role in the events leading up to the riot remains a central topic in legal and political discussions.

11. Refusal to Condemn White Supremacist Groups (2017)

- During a 2017 debate, Trump was asked to condemn white supremacist groups after violence broke out in Charlottesville, Virginia. He stated that there were “very fine people on both sides” and was criticized for not fully denouncing white supremacy.

12. Threats Against Political Opponents

- Trump often made personal attacks on his political opponents. He referred to Senator Elizabeth Warren as "Pocahontas" to mock her claims of Native American ancestry, and he was known for name-calling, such as calling former President Barack Obama "a terrible president" or referring to Hillary Clinton as "crooked."

13. Trump’s Use of Twitter (Pre-2021)

- Trump was known for his unfiltered tweets, many of which were controversial, attacking political opponents, the media, and anyone critical of him. Some of his tweets were labeled as misinformation or incitement to violence, especially those related to the 2020 election.

14. Dismissal of Russian Interference in 2016 Election

- **Despite intelligence agencies concluding that Russia interfered in the 2016 election to help his campaign, Trump repeatedly downplayed or denied Russia's role, even siding with Vladimir Putin over U.S. intelligence officials.**

15. Lack of Condemnation After Capitol Riots

- **Trump's initial response to the January 6th Capitol insurrection was widely seen as inadequate, and he delayed addressing the violence for hours. His refusal to immediately condemn the rioters contributed to the perception that he incited the attack.**

16. Handling of the George Floyd Protests (2020)

- **Trump's response to the protests against police brutality in the wake of George Floyd's death was controversial. He called for the use of military force to quell the protests, and his decision to have protesters cleared from Lafayette Park for a photo op in front of St. John's Church drew widespread criticism.**

17. Support for Conspiracy Theories (2020-2021)

- **Trump gave credence to numerous conspiracy theories, particularly around the 2020 election, where he falsely claimed that the election was "stolen" from him, despite no evidence of widespread fraud.**

18. Hush Money Payments and Stormy Daniels (2016)

- **Trump's alleged affair with porn star Stormy Daniels and the subsequent hush money payments to silence her ahead of the 2016 election became a major legal and political controversy. Trump's involvement in the payment was a central issue in his 2024 legal case.**

19. The Border Wall (2017-2021)

- **Trump's obsession with building a wall on the U.S.-Mexico border became one of his most controversial campaign promises. His administration allocated funds for the wall, and he declared a national emergency in 2019 to secure additional funding after Congress refused to approve it.**

20. Bounty on U.S. Troops (2020)

- **The New York Times reported that Russian intelligence placed bounties on U.S. soldiers in Afghanistan, and Trump was criticized for not confronting Putin about it. He denied the claims and downplayed the issue, despite significant evidence of Russian involvement.**

21. The "Locker Room Talk" Incident

- **In 2005, Trump was recorded talking about his ability to sexually assault women because of his fame. He dismissed these comments as "locker room talk" and apologized, but the remarks led to widespread backlash.**

22. Refusal to Commit to Peaceful Transfer of Power (2020)

- In the final weeks of the 2020 election, Trump refused to commit to a peaceful transfer of power, leading to concerns about potential unrest following the election.

23. Trump's Comments on Climate Change (2017-2020)

- Trump dismissed climate change as a hoax and withdrew the U.S. from the Paris Agreement in 2017. He frequently downplayed the science behind global warming and was criticized for his environmental policies.

24. Rhetoric on Immigrants

- Trump has often used inflammatory language when discussing immigrants. He famously referred to Mexican immigrants as "rapists" and criminals, leading to accusations of racism and xenophobia.

25. Involvement with Jeffrey Epstein (human trafficker, sex slaver and financier)

Trump has known ties to Jeffrey Epstein and visited his island where underage girls were forced into sexual slavery.

26. Conviction of Election Tampering, Business Fraud and Sexual Abuse

Trump has been found guilty in both criminal and civil courts for many felonies.

These are just some of the most significant controversies surrounding Trump. His time in office and his subsequent political career have been marked by numerous moments that have divided public opinion, with some praising his unorthodox approach and others condemning his rhetoric and actions.

PART XXXV – Trump's Relationship with Sex Slave Trafficker Jeffrey Epstein

Donald Trump's relationship with financier and convicted sex offender Jeffrey Epstein has been a subject of significant media attention and scrutiny, particularly following Epstein's arrest in 2019 on federal charges of sex trafficking minors. Their association is marked by both public appearances and a network of high-profile connections, though the details of their personal relationship remain complex and layered.

Early Encounters and Public Appearances

Trump and Epstein reportedly knew each other in the 1990s and early 2000s, both being prominent figures in New York's social scene. Trump was a real estate mogul and frequent figure at elite parties, while Epstein, a financier with a mysterious background, was known for mingling with celebrities, political figures, and business magnates. The two were photographed together at various events, including parties at Mar-a-Lago, Trump's estate in Palm Beach, Florida. In the early 2000s, Trump even praised Epstein, calling him a "terrific guy" and saying that Epstein had a "lot of fun" with women and liked young women.

Trump and Epstein were part of a larger circle of wealthy individuals, and their social interactions seemed to be centered around mutual business interests and a shared affinity for lavish lifestyles. Trump's ties to Epstein were widely acknowledged until Epstein's legal issues became more public in the mid-2000s.

Epstein's Legal Troubles and Trump's Denial

Epstein was first arrested in 2005 on charges of soliciting minors for sex, and in 2008, he reached a controversial plea deal that allowed him to serve just 13 months in a county jail, despite facing serious allegations. Over the years, his association with influential people like Trump remained under scrutiny, and rumors about his illicit activities persisted.

As Epstein's crimes became more publicly known and he was arrested again in 2019, Trump was asked about his relationship with Epstein. He claimed that he had a falling out with Epstein years earlier, though he did not elaborate on the reasons. In a 2019 interview, Trump distanced himself from Epstein, saying he "was not a fan" and "hadn't spoken to him in years." Trump also noted that he was "not a part of that world" and that he had "a lot of respect for the people who were trying to do something about it."

However, the connection between Trump and Epstein continued to generate headlines due to the extensive social circles they shared and Epstein's access to powerful people, including Trump's political allies, former presidents, and royalty.

Trump's Private and Public Interactions with Epstein

There are multiple documented interactions between Trump and Epstein, some of which raise concerns. Notably, in 1992, Trump allegedly kicked Epstein out of Mar-a-Lago after Epstein allegedly made an inappropriate advance towards a young woman. But there are conflicting accounts regarding the nature of their interactions, as Trump has denied any knowledge of Epstein's criminal activities.

Trump has also faced questions about a 2002 interview he gave to *New York Magazine*, in which he said, "I've known Jeff for 15 years. Terrific guy. He's a lot of fun to be with. It is even said that he likes beautiful women as much as I do, and many of them are on the younger side."

Despite this, Trump has consistently downplayed his association with Epstein, denying any involvement in Epstein's illegal activities. He has also been adamant about not being part of the "private jet" or "pedophile" circles that Epstein was accused of being involved in, even though several women have claimed they were trafficked by Epstein and had encounters with high-profile men, including Trump.

The 2016 Election and Ongoing Legal Scrutiny

During the 2016 presidential campaign, Trump's ties to Epstein resurfaced, with media outlets questioning the nature of their relationship. However, Trump's campaign dismissed concerns, with Trump himself referring to Epstein as "a nice guy" in the past, but maintaining that he hadn't seen him in years.

Epstein's death in August 2019, ruled a suicide in federal custody, further complicated matters. His death raised suspicions and conspiracy theories, many involving the powerful people in Epstein's orbit. Some theorists speculated that Trump was one of many figures who might have been involved in Epstein's alleged sex trafficking ring, although no direct evidence has surfaced linking Trump to Epstein's criminal activities.

Legal Implications and Victims' Allegations

Several women who have accused Epstein of sexual abuse (Trump has been found guilty of such sexual abuse of other women) have also mentioned Trump in their testimony, but none have directly accused him of participating in the crimes Epstein allegedly committed. In 2016, a woman named "Jane Doe" filed a lawsuit against Trump and Epstein, claiming she was sexually assaulted by both men when she was 13 years old. The case was eventually dropped, with Trump's legal team calling the accusations "false" and "without merit." Trump has consistently denied any involvement in Epstein's crimes.

In the wake of Epstein's death and the continued legal proceedings involving his associates, including Ghislaine Maxwell (Epstein's close associate and convicted sex trafficker), Trump's ties to Epstein remain a subject of public interest and legal scrutiny. However, Trump has largely evaded legal consequences for his relationship with Epstein, as no direct evidence has emerged that implicates him in Epstein's criminal activities.

In a 2017 interview released as part of the *Fire and Fury* podcast series, journalist Michael Wolff and former NME editor James Truman unveiled audio recordings of Jeffrey Epstein discussing his close social relationship with Donald Trump, a connection Trump had long denied. The tapes, which are part of a larger collection of over 100 hours of Epstein's reflections, reveal Epstein's insights into the dynamics within Trump's inner circle and his management style.

Epstein recalls how Trump played his advisors against each other, sowing discord by spreading rumors and manipulating their relationships for his benefit. He describes how Trump would criticize staff like Steve Bannon, Reince Priebus, and Kellyanne Conway while simultaneously exploiting their weaknesses to maintain control. Epstein paints a picture of Trump as a master manipulator who used division to maintain his grip on power.

Despite Trump's public distancing from Epstein after his 2019 arrest on sex trafficking charges, Wolff's podcast suggests a deeper, long-standing connection between the two. The two men were known to have socialized in elite circles, sharing interests in wealth, women, and status. Wolff compares their bond to one of mutual exploitation, with both men seeking dominance and control.

The release of the tapes also revisits troubling allegations against Trump. One example is a claim from former model Stacey Williams, who alleged that Trump groped her in 1993 at Epstein's urging. Trump has denied this claim.

The podcast sheds light on the dark, intertwined history of two powerful men, both driven by similar desires, one of whom ended up in prison while the other ascended to the presidency. Wolff's aim is to highlight the disturbing parallels between their paths, illustrating how their shared obsession with power and influence led to vastly different fates.

The relationship between Donald Trump and Jeffrey Epstein is a complex and multifaceted topic, characterized by their connections in New York's social elite and the murky waters of Epstein's criminal activities. While Trump has consistently denied involvement in Epstein's alleged trafficking activities, their public interactions and social connections remain part of a broader narrative about Epstein's powerful circle. As Epstein's associates face legal repercussions, questions about the full extent of his relationships with high-profile individuals, including Trump, will likely continue to resurface. Epstein has said that he was Trump's closest friend.

PART XXXVI – Trump's Relationship with Nazi-saluting billionaire oligarch Elon Musk

Donald Trump and Elon Musk's relationship has fluctuated over the years, shifting from mutual admiration to public tension. Both figures, icons in their respective fields—Trump in politics and business, Musk in technology and innovation—share a penchant for disruption and a drive for American greatness, but their paths and philosophies diverge in significant ways.

Early Interactions and Shared Interests

In the early days of their careers, Trump and Musk seemed to occupy separate spheres. Trump's empire, built on real estate and branding, stood in stark contrast to Musk's high-tech ventures with companies like Tesla, SpaceX, and SolarCity. However, both men garnered attention for their bold ambitions, and their early interactions reflected a shared belief in the power of American entrepreneurship.

Musk briefly joined Trump's Strategic and Policy Forum in 2017, which consisted of high-profile business leaders who advised the president on economic policy. However, when Trump decided to pull the U.S. out of the Paris Climate Agreement in June 2017, Musk resigned, citing a disagreement over climate change policies. This moment marked the first real public rift between the two, as Musk advocated for sustainable energy while Trump remained committed to policies that favored traditional industries.

Public Tensions and Diverging Philosophies

While the initial split over the Paris Agreement created a public divide, the two men's paths continued to cross in the years that followed, especially as Trump's administration embraced deregulation and manufacturing in America. Musk, in turn, found common ground with Trump on issues like reducing business taxes and strengthening American innovation. At the same time, Musk's push for green energy and climate action sharply conflicted with Trump's environmental policies, including his focus on reviving the coal industry and rolling back regulations.

Throughout this period, Trump would occasionally praise Musk's accomplishments, acknowledging his role in transforming industries like electric vehicles and space exploration. However, Musk remained skeptical of Trump's approach to environmental and technological challenges, especially as it related to clean energy and government intervention in the economy.

2020 Election and Their Post-Election Dynamic

During the 2020 election, Musk expressed support for some of Trump's policies, particularly those that benefited American businesses and manufacturing. However, Musk remained more neutral in his political stance compared to other figures in Trump's orbit. Musk's endorsement of free-market innovation seemed to clash with the broader political landscape, and his views on government restrictions, especially during the COVID-19 pandemic, mirrored Trump's rhetoric about economic recovery and personal freedoms.

In the aftermath of the 2020 election, tensions simmered again. Musk distanced himself from the political drama surrounding Trump's claims of a stolen election, opting to remain largely silent while Trump's legal challenges gained national attention. Nevertheless, both men remained vocal critics of government regulations and lockdowns, positioning themselves as defenders of personal liberty and economic recovery.

Recent Tensions: Public Feuds and Shifting Allegiances

As of late 2024, their relationship has soured further. In a notable instance, Trump escalated his feud with Musk in a Truth Social post, belittling the tech billionaire. Musk, who had previously been a vocal advocate for some of Trump's business policies, has found himself at odds with the former president more publicly than ever before. The tension stems from various political and business disagreements, with Trump making critical remarks about Musk's leadership and Musk responding with pointed jabs.

Despite these growing animosities, the core of their relationship remains rooted in mutual recognition of each other's entrepreneurial success. Musk has long viewed innovation as essential to American leadership, while Trump continues to advocate for the country's economic resurgence through deregulation and a focus on traditional industries.

2025 Incidents of Nazi Salute, Stealing Social Security and Medicare Money, etc.

Musk has been prominent in all events, actions and the inauguration of Trump. He is deemed by many as a "shadow president" with access to government agencies and personal records of Americans (including federal workers). He has been designated as the head of a fake government agency called the Department of Government Efficiency (D.O.G.E.).

Musk has threatened "war" on MAGA followers during an exchange with Trump's base on X (Twitter, which Musk owns). Musk had anyone who disagreed with him after he insulted American workers and said American jobs should go to workers through visa. Musk then had those he was threatening blocked temporarily on X's platform. Trump may or may not be aware of what Musk is planning and carrying out with his given power (though never elected to office or confirmed by Congress as an appointee to lead a federal agency). There is also the Nazi/ Hitler salute Musk had done twice during Trump's inauguration and his speaking with a German far-right party telling them to get over the guilt of the past (referring to Nazis in Germany during World War II and the Holocaust). As of this writing, Musk has tried to get access to both records and funds of Social Security and Medicare payment systems by bullying the Treasury top executive, David Lebryk, who has served in numerous high-level roles at the Treasury Department since 1989 and was temporarily named acting treasury secretary by Trump before the confirmation this week of his nominee, Scott Bessent, would soon leave the department. The agency distributes more than $6 trillion annually to households and businesses.

It is unknown if Musk was doing this on his own behalf, that of Trump or some other third party like the foreign countries Musk has business dealings with.

PART XXXVII – Trump's Denial of 2016 Election Loss and the Capitol Riot

In the wake of the 2020 presidential election, the United States found itself at a pivotal moment in its democratic history. After months of campaigning, record voter turnout, and an election that saw Joe Biden defeating incumbent Donald Trump by a significant margin, the world awaited the peaceful transfer of power that had long been a hallmark of American democracy.

However, what followed was anything but peaceful. Instead, it marked the beginning of a relentless campaign of misinformation, legal challenges, and an unprecedented attack on the democratic process.

The Beginning of the Denial

From the night of the election, Trump refused to accept the results. Despite early votes showing Biden leading, the counting of mail-in ballots – a result of the COVID-19 pandemic – revealed a shift towards Biden as more votes were tallied. In the early hours of November 4, Trump declared victory, even as the results were far from final. His words were a precursor to the months of false claims that followed.

“Frankly, we did win this election,” Trump stated repeatedly, casting doubt on the legitimacy of mail-in ballots and the voting process as a whole. By the end of the month, Trump had begun filing a series of lawsuits in battleground states. Each was swiftly dismissed or debunked, but the narrative of a "stolen" election persisted.

Trump and his allies focused on unfounded allegations of voter fraud, claiming without evidence that Dominion Voting Systems, voting machines, and local officials had rigged the outcome. Despite widespread consensus from election officials, both Republican and Democrat, that the election was fair, Trump's rhetoric grew more inflammatory. He pressured Republican lawmakers to take action, even urging them to overturn results in swing states.

The "Stop the Steal" Movement

In December, the narrative shifted into overdrive with the creation of the "Stop the Steal" movement. Thousands of Trump supporters, fueled by misinformation, believed the election had been rigged. Trump became increasingly vocal, using his platform to spread disinformation on social media, encouraging his base to gather and protest the results. His supporters amplified the message, and by early January, the rhetoric had morphed into a dangerous call to action.

On January 6, 2021, Congress was set to certify the Electoral College results, officially naming Joe Biden as the next President of the United States. Trump, now emboldened by weeks of denial, organized a rally in Washington, D.C. that would precede the certification process. “We’re going to walk down to the Capitol,” he told his supporters, “and we’re going to cheer on our brave senators and congressmen and women.”

He had no intention of walking to the Capitol himself, but his words had a profound effect on the crowd. They gathered at the National Mall, listening to a series of inflammatory speeches from Trump allies, including Rudy Giuliani and Donald Trump Jr. The rhetoric was clear: they had been cheated, and something needed to be done about it.

The Storming of the Capitol

At approximately 1:00 PM, as the Joint Session of Congress convened to certify the Electoral College results, Trump’s supporters began marching toward the Capitol. The atmosphere was tense, but it was nothing compared to the chaos that would soon unfold.

By 2:00 PM, a violent mob had breached the Capitol's barricades and stormed the building. Lawmakers, staff, and security personnel were forced into lockdown. Windows were smashed, offices were ransacked, and flags – including that of the Confederacy – were hoisted inside the Capitol.

The scene was chaotic and horrifying, with rioters vandalizing the historic halls of American democracy.

For hours, the Capitol was in lockdown as police and National Guard units struggled to regain control. The violence disrupted the certification of the election results, as lawmakers were evacuated to safe locations. Senate Majority Leader Mitch McConnell, Speaker of the House Nancy Pelosi, and others were among those forced to flee.

The breach of the Capitol was unprecedented in American history, an assault on the very institution designed to uphold the will of the people. The images of the mob roaming freely through the halls of Congress, some even sitting at the desks of lawmakers, were broadcast worldwide.

Trump's Response

Throughout the chaos, Trump remained silent. He had been urged to call for peace, but his response was slow and ineffective. He posted a video on Twitter telling the rioters to go home, but added, "We love you, you're very special," sending a mixed message that further enraged critics. He later posted a message calling for peace but continued to assert that the election had been stolen.

His lack of condemnation of the violence was widely criticized. Many Republicans, once loyal to Trump, condemned the events and the role his rhetoric had played in inciting the riot. Among them was Senator Mitch McConnell, who had supported Trump throughout his presidency but now denounced him in the wake of the Capitol attack.

Aftermath and Consequences

By evening, the Capitol was cleared, and lawmakers reconvened to finish certifying the Electoral College vote. Despite the chaos, the certification process was completed in the early hours of January 7, 2021. Joe Biden's victory was officially confirmed.

In the days following the riot, the fallout from Trump's actions was swift. He was impeached for a second time by the House of Representatives, this time for "incitement of insurrection," making him the first president in U.S. history to be impeached twice. While he was acquitted in the Senate, his influence over the Republican Party had been irrevocably damaged.

Social media companies, including Twitter and Facebook, banned Trump from their platforms, citing the risk of further violence and disinformation. In the months that followed, numerous investigations were launched into the events of January 6, and hundreds of rioters were arrested and charged with various crimes.

Trump's refusal to accept the election results, his repeated calls for action, and the violent storming of the Capitol marked a dark chapter in American history. The impact of that day would be felt for years to come, as the nation grappled with the erosion of democratic norms and the deepening political divide.

Reflection

The denial of the 2020 election results and the events of January 6, 2021, brought to light the fragility of democracy in America. For many, it was a wake-up call, a reminder that institutions and norms that had been taken for granted could be challenged by misinformation and extremist rhetoric. It revealed the vulnerabilities in a system that, while resilient, was not immune to manipulation or attack.

As the nation moved forward, questions about accountability, the role of social media, and the future of American democracy remained at the forefront. But one thing was clear: the events of that fateful day would forever be etched in history, not only as a tragic attack on the Capitol but as a warning of the dangers posed when truth is abandoned and political power is used to incite division.

PART XXXVIII – A Cautionary Note (The Rise of Fascism in Nazi Germany)

Adolf Hitler's rise to power and the transformation of Germany into a fascist state is one of the most significant events of the 20th century. His ascent and the subsequent establishment of a totalitarian regime under the Nazi Party had far-reaching consequences, not only for Germany but for the world. This chapter will examine how Hitler gained power, the key events that allowed him to solidify his control, and the steps he took to turn Germany into a fascist state.

1. The Weimar Republic and the Conditions for Hitler's Rise

After Germany's defeat in World War I, the country was left politically and economically unstable. The Weimar Republic, established in 1919, was marked by a fragile democracy, which faced numerous challenges. The Treaty of Versailles (1919) imposed heavy reparations on Germany, causing severe economic hardship, which was exacerbated by the global Great Depression in 1929. The people of Germany, embittered by their defeat and the harsh terms of the treaty, were disillusioned with the Weimar government. The republic's inability to address the economic crisis and the growing sense of humiliation among the population created a fertile ground for extremist political movements.

The Nazi Party, led by Adolf Hitler, began to exploit these grievances. Hitler's powerful oratory skills, his nationalistic rhetoric, and promises to restore Germany to its former glory attracted a growing number of followers.

2. The Nazi Party's Strategy and the Road to Power

Hitler became the leader of the National Socialist German Workers' Party (NSDAP), or Nazi Party, in 1921. He skillfully used propaganda to appeal to various segments of society, including disaffected veterans, middle-class citizens worried about communism, and nationalists who wanted to restore Germany's power.

In the early 1930s, as the Great Depression worsened, the Nazi Party's popularity surged. The party's platform, which included rejecting the Treaty of Versailles, advocating for the rearmament of Germany, and promoting anti-Semitism, resonated with many Germans who were struggling in the economic crisis. The Nazis also promised to bring order to the country, which was appealing in a time of widespread social unrest.

In the 1932 elections, the Nazi Party became the largest party in the Reichstag (German parliament), though they did not have an outright majority. Hitler's political maneuvering and backroom deals played a crucial role in securing his position. Ultimately, on January 30, 1933, President Paul von Hindenburg appointed Hitler as Chancellor of Germany in a coalition government.

3. The Reichstag Fire and the Reichstag Fire Decree

After Hitler was appointed Chancellor, he moved quickly to consolidate his power. One of the first key events that helped him was the Reichstag Fire, which occurred on February 27, 1933. The fire, which destroyed the Reichstag building, was blamed on a Dutch communist named Marinus van der Lubbe, although the true cause remains disputed.

Hitler and his allies seized on the fire as an opportunity to push through emergency measures. The Reichstag Fire Decree, issued the following day, suspended civil liberties, including freedom of speech, assembly, and the press. It also allowed for the arrest of political opponents, especially communists, who were seen as the primary threat to the Nazi regime. The decree effectively eliminated political opposition and helped Hitler secure a firmer grip on power.

4. The Enabling Act and the End of Democracy

On March 23, 1933, the Reichstag passed the Enabling Act, which gave Hitler the authority to enact laws without the approval of the Reichstag or President Hindenburg. This effectively gave Hitler dictatorial powers and marked the end of the Weimar Republic's democratic government.

The Enabling Act was passed with the support of the Nazi Party and their allies, and it was approved by a large majority in the Reichstag, although many communists and socialists had already been arrested or intimidated. With this law, Hitler no longer had to rely on parliamentary support to govern and could make decisions by decree. This marked the beginning of the establishment of a totalitarian state.

5. The Purge of Political Opponents: The Night of the Long Knives

Once in power, Hitler moved to eliminate any remaining potential threats to his authority, even within his own party. In June 1934, Hitler ordered a purge of the SA (Sturmabteilung), the Nazi paramilitary wing, which had grown too powerful for Hitler's comfort. The SA, led by Ernst Röhm, had played a crucial role in the Nazi rise to power, but Röhm and others in the organization had become a threat to Hitler's control.

On the night of June 30, 1934, known as the Night of the Long Knives, the SS (Schutzstaffel), another Nazi paramilitary group, carried out a series of murders of SA leaders, as well as political opponents and rivals within the Nazi Party. The purge solidified Hitler's control over the Nazi Party and eliminated internal threats to his rule.

6. The Gleichschaltung (Coordination) and the Nazi Totalitarian State

Following the passing of the Enabling Act and the elimination of political opposition, Hitler and the Nazi Party began a process known as Gleichschaltung, or "coordination," in which all aspects of German society were brought under Nazi control. The goal was to create a totalitarian state in which the Nazi Party controlled every element of life in Germany.

The Nazis established a single-party state, and political opposition was banned. Trade unions were dissolved and replaced by the German Labor Front, a Nazi-controlled organization. All media, including newspapers, radio, and films, were tightly controlled by the state. Education was restructured to teach Nazi ideology, and youth organizations, such as the Hitler Youth, were created to indoctrinate children into Nazi beliefs.

Religious institutions were also targeted. While Hitler did not fully eradicate religion, he sought to control and manipulate it. The regime signed a concordat with the Catholic Church in 1933, but over time, it undermined religious autonomy and attempted to replace Christianity with a state-sponsored version of the Nazi ideology.

7. The Cult of Personality and the Führer Principle

As Hitler consolidated power, he developed a personality cult, positioning himself as the infallible leader or "Führer" of the German people. The Führer principle became central to the Nazi state: Hitler was portrayed as the ultimate authority, with complete control over the government, military, and society. The image of Hitler as the strong, decisive leader was central to Nazi propaganda, and his will was presented as the will of the nation itself.

8. The Nazi Economy and Rearmament

To maintain popular support and strengthen his position, Hitler initiated large-scale public works programs to address the economic crisis and reduce unemployment. The construction of the Autobahn (highways) and other infrastructure projects provided jobs and boosted the economy.

However, one of the key goals of Hitler's regime was the rearmament of Germany, in violation of the Treaty of Versailles. The Nazi regime poured resources into building a powerful military and expanding Germany's war potential. This militarization of the economy not only helped end the Great Depression in Germany but also set the stage for the aggressive expansionism that would lead to World War II.

9. The Final Transformation into a Fascist State

By the mid-1930s, Germany had fully transformed into a fascist state. The Nazi Party, under Hitler's leadership, had complete control over every aspect of life in Germany. Political dissent was crushed, opposition groups were eliminated, and all institutions were aligned with Nazi ideology. The regime was characterized by extreme nationalism, militarism, anti-Semitism, and the rejection of democratic principles.

Hitler's rise to power and the establishment of the Nazi state was marked by violence, intimidation, and manipulation. The regime rapidly became one of the most brutal totalitarian systems in history, leading to the persecution of millions of people, especially Jews, and eventually to the outbreak of World War II.

In conclusion, Hitler's path to power was a combination of strategic political maneuvering, exploitation of economic and social unrest, and ruthless elimination of opposition. Once in power, he swiftly dismantled democratic institutions and transformed Germany into a fascist dictatorship. This totalitarian regime set the stage for the horrors of the Holocaust and the devastation of World War II, leaving a profound and lasting impact on world history.

PART XXXIX – Best and Worst Case Scenarios for Trump's Second Presidential Term

The best case scenario would be Trump helping every citizen reach the American Dream.

With that being said, it is statistically impossible that to be our reality with all of this analyzed data in the A.I. predictive models. Based on the analyzed data that has been presented in previous parts of this book, Artificial Intelligence has predicted this through its models.

Let us go over how fascism could rise to take over a democracy like the United States of America.

To discuss how Donald Trump or any leader could potentially turn a democratic nation into a fascist state, we can look at historical examples and modern political dynamics. Fascism is generally defined as a far-right authoritarian ultranationalist ideology, often characterized by dictatorial power, forcible suppression of opposition, strong regimentation of society, and control over many aspects of life, including the economy and culture. Here are several ways in which a leader like Trump could theoretically work towards turning the United States into a fascist state:

1. Consolidation of Power

Fascist regimes often centralize power in the hands of one leader or a small elite group. Trump's use of executive orders and his often combative relationship with other branches of government during his presidency hinted at a willingness to bypass democratic institutions. For example, his attempts to delegitimize the judicial system and Congress, particularly when he disagreed with them, raised concerns about the undermining of checks and balances. If a future leader were to continue or escalate these tactics, they could dismantle the democratic system and consolidate authority in the executive branch.

2. Manipulation of Public Opinion

Fascist leaders often rely heavily on propaganda and media control to shape public perception and reduce dissent. Trump's frequent attacks on the media, calling it "fake news" or labeling it as the "enemy of the people," could contribute to the undermining of a free press. By promoting alternative narratives or creating a parallel media ecosystem, a leader like Trump could manipulate public opinion and suppress dissent. This is an essential strategy in fascist states, where the ruling party controls the narrative and suppresses opposition.

3. Erosion of Democratic Norms and Institutions

Fascist movements often erode the integrity of democratic institutions, such as the courts, the press, or the electoral system. Trump's refusal to concede the 2020 election, his efforts to challenge the election results without evidence, and the January 6th Capitol riot are examples of how democratic norms can be weakened. If a future leader continues to undermine the electoral process, delegitimize opposition parties, or refuse to recognize the outcome of elections, it could further erode trust in democratic institutions, paving the way for authoritarian rule.

4. Targeting and Demonizing Specific Groups

A hallmark of fascist regimes is the identification of a "common enemy" to unite the population against, often through the demonization of minority groups. Trump frequently targeted immigrant communities, Muslim Americans, and Black Lives Matter protesters, which polarized the nation. While these attacks are not new in history, they can create the conditions for scapegoating, violence, and systematic discrimination. A leader who continues to inflame racial, ethnic, or political divisions could move a society closer to fascism.

5. Militarization of Society

Fascist regimes often increase the militarization of society, either through a strong military presence or the creation of paramilitary groups. Trump's rhetoric often supported police and military forces, including during protests, and in some cases, he praised or emboldened far-right groups like the

Proud Boys. These groups, if further empowered, could act as enforcers of a fascist agenda, intimidating and suppressing opposition both legally and illegally.

6. Undermining Civil Liberties

Fascist regimes suppress civil liberties, often through surveillance, censorship, and repression of political opposition. During Trump's tenure, there were concerns about increasing surveillance powers, the expansion of executive authority, and the use of law enforcement for political purposes. If such tactics were further normalized, civil liberties could be severely restricted, and opposition groups could face criminalization.

7. Fostering Nationalism and Isolationism

A critical element of fascism is the promotion of extreme nationalism or patriotism. Trump often employed nationalistic rhetoric, focusing on "America First" policies and appealing to populist sentiments. Nationalism can be a tool for creating a unified, us-versus-them mentality, which can help justify the exclusion or persecution of certain groups and pave the way for authoritarian rule. His tariff war is an example of isolationism.

8. Using Legal and Constitutional Manipulation

While fascist movements often rise through illegal or unconstitutional means, they can also work within the legal framework to make themselves more entrenched. Trump showed interest in changing legal and constitutional structures to serve his agenda, such as calling for changes to voting laws or proposing the disqualification of certain political opponents. If such actions are pursued aggressively, it could dismantle key legal protections that preserve democracy.

Conclusion

Turning a democracy into a fascist state involves a gradual erosion of democratic norms, the centralization of power through unchecked power grabs, the creation of an atmosphere of fear and division, and the eventual establishment of authoritarian rule. While there is no certainty that Trump or any other leader would take all these steps, some of the actions that he took during his time in office – and that he has continued to champion afterward – provide a roadmap for how fascist ideologies could emerge. However, it's important to note that the US has strong institutions, checks and balances, and a vibrant civil society that can resist such movements. Fascism is a danger to any democracy, but it's also a challenge that many democratic societies, including the United States, must remain vigilant against.

The worst-case scenario under a second Trump administration could involve the following:

1. Use of Violent Force to Quell Opposition: Trump has historically suggested using heavy-handed tactics to suppress protests, particularly with his endorsement of law enforcement using force against demonstrators. His rhetoric during the 2020 protests and his attempts to mobilize military forces against American citizens suggest a willingness to escalate force, including against perceived political enemies.
2. Mass Deportations and Detention: Trump's immigration policies, such as his "zero tolerance" stance, led to family separations and mass deportations of undocumented

immigrants. Under Project 2025, his administration could ramp up these actions not just against illegal immigrants, but potentially targeting political dissidents, activists, or anyone opposing his agenda, under the guise of "law and order."

3. Executive Overreach: Trump has already shown a tendency to issue sweeping executive orders to circumvent Congress, like his travel bans and attempts to dismantle the Affordable Care Act. In a second term, these actions could escalate, with Trump issuing even more controversial executive orders, bypassing legal and ethical standards, such as his past calls to limit free speech on social media and media outlets that criticize him.

4. Consolidating Power and Neutralizing Opposition: Through his allies and efforts to pack the courts with conservative judges, Trump could further undermine the checks and balances of the federal government. His idea of controlling both the legislative and judicial branches would allow him to pass laws unchecked, potentially eliminating or manipulating the role of Congress and the Supreme Court to further entrench his rule.

5. Criminalizing Dissent: Trump has already made several attempts to silence critics, including threats to “loosen” libel laws against journalists. Project 2025 could include more aggressive tactics, such as criminalizing speech that contradicts his views or actions, including media outlets, educational institutions, and social media platforms.

6. Dismantling Federal Agencies and Public Services: Trump’s deconstruction of federal agencies, such as the Environmental Protection Agency and the Department of Education, could worsen under Project 2025, leaving essential services vulnerable. The gutting of aviation safety regulations and other watchdog agencies under Trump and his allies, like Elon Musk’s influence over the aviation industry, demonstrates a troubling disregard for public safety, leading to catastrophic consequences like crashes, fatalities, and infrastructure failures.

7. Financial Exploitation and Corruption: Trump has been accused of using his presidency to enrich himself, from profiting off his hotels to seeking foreign business deals. His repeated attempts to undermine federal funding and the public financial system raise concerns that under Project 2025, he and his associates could engage in further financial malfeasance, potentially siphoning off trillions from the U.S. economy.

8. Targeting Minority Groups: Trump's history of stoking racial tensions, including his attacks on affirmative action and initiatives designed to promote diversity, equity, and inclusion (DEI), could intensify in a second term. Project 2025 may see a further erosion of protections for marginalized groups, leading to more systemic harm against communities of color, immigrants, LGBTQ+ individuals, and others, particularly if linked to policies that scapegoat them for social or economic problems.

9. Empowering Extremist Groups: Trump has been criticized for his relationship with hate groups, including white nationalists and violent extremists. His rhetoric and actions have emboldened these groups, as seen on January 6th, 2021, where his supporters attempted an insurrection. In a second term, Trump could further legitimize these groups, either by pardoning insurrectionists or turning a blind eye to their criminal acts.

10. Aligning with Authoritarian Regimes: Trump's friendly relations with Vladimir Putin and other authoritarian leaders, like Kim Jong-un, point to a potential foreign policy shift that undermines U.S. interests. By allowing foreign powers to exert influence, Trump could weaken national security and foster instability both at home and abroad.

11. Facilitating Oligarchic Control: Trump's policies have often favored the ultra-wealthy and large corporations. Project 2025 could accelerate this trend, leading to a consolidation of power in the hands of a few billionaires, which could further

erode democratic institutions, allowing wealthy elites to control politics and the economy at the expense of the public.

12. Eliminating Democratic Processes: Trump has suggested he may not honor election results that don't favor him, as seen after the 2020 election. Under Project 2025, Trump could push to permanently undermine democracy by halting elections or invoking national emergencies to extend his presidency indefinitely, giving him near-dictatorial powers.

13. Destabilizing the U.S. and the World: Trump's economic and foreign policies, particularly his approach to trade, climate change, and international relations, have already strained the global order. If his policies lead to economic collapse, social unrest, or weakened alliances, it could not only devastate America's future but create global instability, with far-reaching consequences for international security and prosperity.

The nightmare scenario is Trump setting off World War III and using nuclear weapons and completely destroying the United States, if not the world.

A world leader could potentially ignite World War III through a combination of aggressive policies, strategic miscalculations, and an escalating series of global crises. While it's impossible to predict the exact circumstances, history offers several lessons about the factors that could lead to such a catastrophic conflict. Here's how a world leader could contribute to the start of a global war:

1. Aggressive Military Expansion

A leader with expansionist ambitions could invade or annex neighboring countries, much like Adolf Hitler did with Poland in 1939. If a leader in a powerful country (for example,

Russia, China, or the U.S.) were to invade another nation, this could quickly escalate into a global conflict if other countries, particularly alliances like NATO, were drawn into the fight. The global interconnectedness of economies and alliances means that a regional war could easily spiral out of control.

For instance, if a leader decided to militarily invade Taiwan or push for a military confrontation over disputed territories (like in the South China Sea), the United States and other countries with interests in the region could be forced to respond. The escalation of such tensions could lead to a world war. Trump is trying to make Canada the 51st state of the U.S., annexing Greenland from Denmark and seizing the Panama Canal.

2. Nuclear Proliferation or Use

One of the most direct ways to trigger World War III would be through the use of nuclear weapons, whether in an attack or as a result of nuclear escalation. A leader who engages in reckless rhetoric or decisions regarding the use of nuclear weapons could lead to devastating consequences. For example, a leader who threatens to use nuclear weapons as a bargaining tool or conducts a nuclear test near other countries could provoke a military response from those nations, especially if they feel their own national security is at risk.

If a regional conflict involving nuclear powers escalated without de-escalation mechanisms, this could rapidly spread across borders. The proliferation of nuclear weapons, with more countries gaining access to these weapons, could also increase the chances of a nuclear conflict, potentially triggering a global war.

3. Destabilizing Alliances and Treaties

Global peace has been maintained in large part due to international treaties, alliances, and organizations like the United Nations, NATO, or the European Union. A leader who actively undermines or withdraws from international agreements could destabilize global security. For example, withdrawing from arms control treaties, such as the Intermediate-Range Nuclear Forces (INF) Treaty or the Iran nuclear deal, could trigger an arms race, heightening tensions between global powers.

If a world leader breaks these treaties and no diplomatic solutions are found, other countries might be compelled to form counter-alliances, setting the stage for a larger confrontation.

4. Escalation of Proxy Wars

World War III could also be sparked by a conflict that initially starts as a "proxy war" between two superpowers or major powers, where both sides support opposing factions in a third country. A world leader could intensify existing proxy wars, such as those in the Middle East, Eastern Europe, or Africa, and inadvertently drag major powers into direct conflict. For example, involvement in the civil war in Syria or the conflict in Ukraine could lead to an escalation where military actions inadvertently trigger a wider war.

Proxy wars often provide an opportunity for superpowers to fight without direct confrontation, but as resources, troops, and armaments are supplied to opposing factions, the risk of direct clashes increases.

5. Economic Warfare and Trade Disputes

While not traditionally viewed as a direct route to war, economic warfare can lead to conflict, especially if one country tries to cripple another's economy through sanctions or trade blockades. A leader who escalates economic disputes—especially over resources like oil, technology, or critical infrastructure—could strain international relations and lead to military conflict. If a country's economy is severely impacted, it might turn to war as a means of defense, and if major economic powers engage in a trade war or aggressive sanctions, the domino effect could pull other nations into conflict.

For instance, if the U.S. were to impose crippling sanctions on China or another rising power, China might respond with military action, which could lead to direct confrontation with the U.S. and its allies.

6. Nationalism and Rhetoric of Revenge

Leaders who promote aggressive nationalism, historical grievances, or xenophobia could heighten tensions within and between nations. This kind of rhetoric can stir up ethnic or national conflicts and be used as a justification for war. A leader who calls for the reclaiming of territories or the revenge of past wrongs (e.g., a country seeking to avenge a perceived historical humiliation) could rally domestic support, but also provoke other nations, particularly if the territorial claims involve disputed regions or allies of other major powers. The rhetoric of military expansion or historical revenge is dangerous because it can rally domestic nationalism but ignite regional or global tensions. The path from words to action in such cases can be dangerously short.

7. Cyber Warfare and Escalation

With the rise of cyberwarfare, leaders now have the ability to target critical infrastructure, disrupt economies, and even disrupt military operations. A cyberattack on a nation's infrastructure—like power grids, financial systems, or military communications—could be perceived as an act of war, especially if the attack originates from another country. If a world leader were to conduct or support large-scale cyberattacks on adversaries, it could escalate tensions quickly, leading to military retaliation and eventually an all-out war.

8. Miscommunication or Human Error

A world leader could, through miscommunication or human error, trigger a war. Military personnel, diplomats, or intelligence officers may misinterpret the actions or intentions of another nation, leading to unintended conflict. An accidental military engagement, like a border skirmish that escalates into a larger confrontation, could easily spiral out of control if both sides are prepared for war and sensitive to perceived threats. A leader who ignores the warnings of diplomacy, or fails to communicate properly with international counterparts, could contribute to such an outcome.

9. Instability in a Key Region

Regional instability in crucial areas—like the Korean Peninsula, the South China Sea, or Eastern Europe—could trigger a global war. A leader who is involved in inflaming tensions in these areas could spark a confrontation that draws in global powers. For example, if a world leader pushed for military confrontation with North Korea or took aggressive actions in the Taiwan Strait, the U.S., China, and their respective allies could be drawn into direct conflict.

Conclusion

World War III could be triggered by a world leader who acts recklessly, either through direct military aggression, manipulation of global politics, or the failure to de-escalate conflicts. The factors that could lead to global war today are complex, involving not just traditional military actions but economic, technological, and diplomatic dimensions. In an interconnected world, even small actions or miscalculations can have far-reaching consequences, and a single leader's decisions could set the stage for global disaster.

However, modern international institutions, diplomatic channels, and the sheer destructiveness of modern weapons make it less likely, though not impossible, for a large-scale war to break out in the traditional sense. Many countries are also aware of the catastrophic consequences of such a conflict and are thus deeply invested in maintaining peace.

To analyze how a leader like Donald Trump could potentially contribute to the outbreak of World War III based on factors like his criminal history, cognitive challenges, and the so-called *Project 2025* plans (which aim to set out a vision for Trump's second presidential term), we need to look at the intersection of his leadership style, personal history, and political ambitions. While there is no concrete evidence that Trump has directly intended to cause global conflict, some elements of his tenure in office and his political proposals could present risks.

1. Criminal History and Leadership Implications

Trump’s criminal history includes a range of allegations, ranging from financial crimes and tax evasion to charges related to the January 6th Capitol insurrection and potential obstruction of justice. While these crimes are not directly linked to causing World War III, a leader with a criminal record could face immense pressures, both domestically and internationally. Such pressure could lead to rash decisions, including:

- Internal Scapegoating and Distraction Tactics: A leader under intense legal pressure may use foreign conflict as a means to rally national support or distract from domestic crises. Leaders in authoritarian regimes often escalate foreign tensions when facing internal threats. If Trump were to face legal consequences, he might increase military or diplomatic confrontations as a way of consolidating power and distracting from scandals.
- Unpredictability in Diplomacy: Trump's legal entanglements and past actions, such as his tendency to challenge norms, could create an unpredictable foreign policy. World leaders may struggle to gauge his intentions, which could lead to miscalculations or escalations. For example, Trump's support for withdrawing from international treaties (such as the Iran Nuclear Deal) was seen as destabilizing, and a return to such policies might strain relations with countries like Iran or North Korea.

2. Cognitive Challenges and Decision-Making

While concerns about Trump's cognitive abilities have been a topic of debate, particularly due to his public speaking and behavior, cognitive impairments (or perceptions of them) could increase the risk of rash, impulsive decisions. A few ways this could manifest include:

- Erratic Foreign Policy Decisions: If a leader shows signs of cognitive decline or appears unpredictable, their adversaries may underestimate them or misinterpret their actions. This could lead to a miscalculation of intentions, especially in high-stakes international situations. Trump's tendency to speak off-the-cuff and make impulsive remarks could escalate tensions with countries like China, Russia, or Iran, especially if there were concerns about his judgment.

- Irrational Responses to Crisis: Cognitive decline could also affect a leader's ability to respond effectively in a crisis, whether that's a military confrontation, an economic collapse, or a diplomatic breakdown. A disjointed or ill-advised response to a diplomatic or military incident could easily snowball into a larger conflict.
- Difficulty in Managing Complex Global Relations: Diplomacy often requires subtlety, long-term planning, and attention to detail. If cognitive impairments interfere with this process, Trump's handling of global crises could lack the nuance necessary to de-escalate dangerous situations, such as those involving nuclear-armed nations or volatile regions.

3. Project 2025 and Nationalist Ambitions

Project 2025 is a set of plans outlined by Trump's team for his second term, focusing on reshaping the executive branch, consolidating power, and implementing far-right policies. There are several aspects of *Project 2025* that could contribute to a more aggressive, militarized, and divisive America, potentially increasing the risks of conflict:

- Erosion of Diplomatic Institutions: *Project 2025* involves the reshaping of the U.S. federal government and its agencies, which could lead to the dismantling of important diplomatic and defense institutions that help maintain global stability. A push toward isolationism, a reduction in international cooperation, and a focus on "America First" could alienate allies and weaken multilateral efforts to prevent war. This type of thinking could make global collaboration to prevent war (especially on issues like climate change or nuclear disarmament) less effective.

- Increased Militarization and Nationalism: *Project 2025* includes a focus on rebuilding and enhancing the U.S. military, as well as reasserting national pride. This could lead to a more belligerent foreign policy stance, especially if paired with aggressive nationalism. If Trump pursued military expansion or interventionist policies under the guise of "national interest," it could heighten tensions with countries like China, Russia, or Iran, leading to the potential for conflict.
- Targeting Political Opponents and Creating Divisions: A leader focused on consolidating power and targeting political opponents could further divide the country. Trump's previous rhetoric and actions have been divisive, and an approach that includes further demonizing domestic enemies could also lead to international instability. Allies may question the U.S.'s reliability, and adversaries may exploit these domestic divisions. This could lead to situations where foreign leaders take advantage of perceived weakness or internal conflict in the U.S., which could increase the risk of international confrontation.

4. Risk of War through Impulsive Military Action

Trump's tendency toward impulsive decision-making—such as the 2019 decision to pull U.S. troops out of Syria, or his reckless rhetoric toward North Korea—raises the possibility of initiating conflict by accident or misjudgment. For example:

- Engaging in Military Provocations: Trump's unpredictable behavior during his first term included aggressive rhetoric against Iran, North Korea, and other adversaries. If a similar approach were taken in a second term, especially with a weakened diplomatic apparatus, it could lead to an inadvertent military clash.
- Potential Military Escalation: Trump's focus on "strength" and military power could lead him to take bold, unprovoked military actions, especially if he perceives a foreign adversary

as weak or vulnerable. This could especially be the case with ongoing tensions in the South China Sea, Taiwan, or Ukraine, where military conflict could trigger broader global repercussions.

- Inability to De-Escalate: If Trump refused to back down during military confrontations or conflicts (such as the ongoing situation in Ukraine or tensions with Iran), it could lead to an irreversible escalation. His combative style, combined with possible cognitive decline, might hinder his ability to reassess situations, leading to further provocations or a global conflict.

Conclusion

While Donald Trump does not appear to have directly intended to cause global conflict, the combination of his criminal history, cognitive issues, and aggressive political plans could increase the risk of geopolitical instability. The unpredictability, nationalism, and military focus of *Project 2025*, along with a potential inclination toward impulsive actions in foreign policy, could create the perfect storm for conflict if tensions with other world powers (like Russia, China, or Iran) escalate into war.

Though much of this is speculative, these elements of his leadership style suggest that under certain conditions, such as a crisis or extreme provocation, his decisions could inadvertently contribute to a broader international conflict. However, it's also important to recognize that the structures of modern global diplomacy and the deterrence provided by nuclear weapons, alliances, and international norms would likely still play a stabilizing role in preventing a full-scale world war.

Regarding nuclear fallout, the prediction is uncertain.

When the A.I. was prompt engineered to predict what Trump would do if given the same playbook as Hitler's Nazi Third Reich in the 1930s-1940s, the prediction was:

Based on historical patterns of authoritarian regimes, including those led by figures like Adolf Hitler, and how those methods could potentially align with the proposals outlined in *Project 2025*. If Trump were to take a more authoritarian or fascist approach in his second term, he might adopt tactics that echo the rise of authoritarian leaders in the past. Here's how that could unfold over the next four years:

Year 1: Rapid Consolidation of Power

1. Suppression of Political Opposition: Trump would likely begin his second term by silencing political opponents, using methods similar to those employed by fascist regimes in the 20th century. In the first few days, he could declare a "state of emergency," justifying the suspension of civil liberties and the concentration of power in the executive branch.

- Political purges: Following a pattern seen in the early years of Hitler's rule, Trump might initiate a sweeping purge of his political enemies, targeting both within his party and the opposition. This could involve imprisoning or disqualifying key political figures and activists through fabricated charges or by utilizing the legal system to discredit them.

2. Legal and Institutional Control: Trump would likely expand the powers of the presidency by utilizing executive orders to directly control the judiciary and law enforcement, following in the footsteps of fascist leaders like Mussolini and Hitler

who consolidated power by taking control of legal and judicial systems. His moves could include:

- Replacing independent judges with loyalists who could rule in his favor, removing checks on his authority.
- Censoring the media: In the manner of fascist regimes, Trump would likely move to control or intimidate the press, turning state-run media or media loyal to his views into the primary source of news.

3. Manipulation of Public Sentiment and Propaganda: Trump might take direct control of media outlets to use propaganda as a tool to solidify his rule. The strategy would be similar to Hitler's use of the Reich Ministry of Public Enlightenment and Propaganda. Through state-run or heavily influenced media outlets, Trump could push his narrative while demonizing his critics and painting his regime as the solution to societal problems.

Year 2: Escalating Attacks on Civil Liberties and the Legal System

4. Militarization and Control Over the Military: In line with authoritarian regimes, Trump would likely seek to use the military as an extension of his power, possibly demanding a loyalty pledge from military leaders to ensure that they are personally loyal to him. The use of the military for domestic purposes could become commonplace, including the suppression of protests and public demonstrations. Trump might even issue orders to suppress political opponents or "enemies of the state" by militarized police forces, much like fascist regimes utilized paramilitary forces to enforce their will.

5. Expansion of Surveillance and Political Repression: Trump would likely adopt methods of surveillance and policing similar to those used by fascist regimes. This could include:

- **Warrantless surveillance of activists and political dissidents, using federal agencies like the FBI and DHS to monitor and intimidate individuals and groups opposed to his government.**
- **Political prisoners: Following the authoritarian playbook, Trump could start imprisoning opposition figures on trumped-up charges, potentially using the justice system to punish political enemies.**

Year 3: State Control and Purges

6. Intensification of the Cult of Personality: Trump would likely ramp up efforts to build a cult of personality, modeling himself as a figure above reproach and central to national identity, similar to Hitler's attempts to become the embodiment of the German people. Statues, monuments, and media campaigns might glorify Trump's legacy and personal rule, while critics would be framed as traitors to the nation.

7. Escalating Violence Against Minority Groups: If Trump fully embraced fascist methods, he might turn a blind eye to or even encourage violence against marginalized groups, in the same way that authoritarian regimes have scapegoated minorities in the past. While *Project 2025* doesn't explicitly call for this, Trump could leverage xenophobia and racism to justify violent repression. For example:

- **Targeting immigrant communities: In a more extreme iteration, Trump could justify violent crackdowns on immigrants, positioning them as threats to the country.**
- **Inciting violence against political opponents: Similar to how Nazi paramilitary groups like the SA and SS were used to attack opponents, Trump could use loyalist groups to engage**

in violent actions against left-wing activists or groups that oppose his rule.

8. Expanding State Control Over the Economy: Trump's efforts to reshape the economy would likely intensify, moving toward a corporatist system where large corporations are directly aligned with state power. This could involve:

- Corporate loyalty tests: Trump could demand that corporations and business leaders pledge loyalty to his regime, similar to how Nazi Germany incorporated big business into the state's agenda.
- State intervention in key industries: Trump might nationalize certain industries to ensure that they serve the interests of the state and his supporters, following a fascist model of economic control.

Year 4: Nationalism and Authoritarianism at Full Force

9. Nationalism and Militarization: Trump's administration would likely push the country into a period of hyper-nationalism, possibly rewriting educational curricula to emphasize his version of American history. This might come with a strong militaristic agenda, focusing on American strength and dominance both at home and abroad. *Project 2025* proposals that call for ramping up military spending would be aligned with this vision.

- Rewriting History: Textbooks and educational systems could be overhauled to glorify Trump's policies and downplay critical views of his administration. Like fascist regimes in the past, he might eliminate "unapproved" views from history.

10. Suppression of Dissent: By the end of Trump's second term, dissent would likely be met with severe retribution. Trump might push to:

- Ban opposition parties: Just as Hitler's regime outlawed opposition parties, Trump might seek to dissolve or neutralize parties or movements that challenge his power.
- Persecution of journalists: Journalists who criticize Trump would likely face legal and physical threats, possibly being jailed or silenced through direct action.

11. Cultivating a War Mentality: Trump's foreign policy might become more aggressive, utilizing the idea of an external threat to solidify his internal control. This could include:

- Escalating trade wars: Engaging in economic and even military confrontations with countries like China, Mexico, or Canada, justifying the militarization of the U.S. and positioning himself as the defender of American sovereignty against foreign invaders.
- Sponsoring paramilitary groups: Just as fascist regimes used unofficial armed groups to maintain order, Trump might encourage the rise of nationalist militias or paramilitary forces to intimidate domestic enemies.

The Final Years: Totalitarian Rule and Unchallenged Power

If Trump successfully implemented all of the *Project 2025* proposals in the manner of past authoritarian regimes, the U.S. could enter a period of totalitarian rule. Political opposition would likely be eradicated, freedom of speech and assembly would be severely curtailed, and dissent would be met with swift punishment. A deep sense of fear and division would likely permeate American society, with some embracing Trump's leadership and others resisting it at great personal risk.

Trump's second term would likely be remembered as a time of extreme political and cultural polarization, with a dramatic shift toward authoritarian governance and the erosion of democratic norms. The consequences would resonate for generations, possibly leading to an eventual reckoning with the authoritarian policies that dominated American life.

PART XL – The Last Word: Donald J. Trump Quotes

In lieu of an official conclusion (being this is an ongoing event), the words of Trump himself may give insight into what the next four years may look like without aid of Artificial Intelligence. There were many more citations, but these offer the most unhindered access to the mind of the president who is causing a constitutional crisis by usurping the power and purse of Congress, by filling his administration with white supremacists, conspiracy theorists and those with links and sympathies for foreign adversaries like Putin and Assad while letting a fake government agency (D.O.G.E.) "run" by a Nazi-saluting Musk access and power of all vital American systems and citizens' personal information.

"… don't let the brevity of these passages prevent you from savoring the profundity of the advice you are about to receive." (How to Get Rich, 2004)

"I am a really smart guy." (TIME, April 14, 2011)

"I'm intelligent. Some people would say I'm very, very, very intelligent." (Fortune, April 3, 2000)

"I know what sells and I know what people want." (Playboy, March 1990)

"I have a great relationship with the blacks. I've always had a great relationship with the blacks." (Albany's Talk 1300, April 14, 2011)

"I just have great respect for them, and you know they like me." (CNN, July 23, 2015)

"A well-educated black has a tremendous advantage over a well-educated white in terms of the job market. ... [I]f I were starting off today, I would love to be a well-educated black, because I believe they do have an actual advantage." (NBC News, September 1989)

"Our great African American President hasn't exactly had a positive impact on the thugs who are so happily and openly destroying Baltimore!" (Twitter, April 28, 2015)

"I have black guys counting my money. ... I hate it. The only guys I want counting my money are short guys that wear yarmulkes all day." (USA Today, May 20, 1991)

"The concept of global warming was created by and for the Chinese in order to make U.S. manufacturing non-competitive." (Twitter, Nov. 6, 2012)

"I know the Chinese. I've made a lot of money with the Chinese. I understand the Chinese mind." (Xinhua, April 2011)

"I did very well with Chinese people. Very well. Believe me." (TIME, April 14, 2011)

"Who the fuck knows? I mean, really, who knows how much the Japs will pay for Manhattan property these days?" (TIME, January 1989)

"The Mexican government forces many bad people into our country. Because they're smart. They're smarter than our leaders." (NBC News, July 8, 2015)

"Jeb Bush will not be able to negotiate against Mexico. Jeb Bush with Mexico said, 'People, come in,' they come in, it's an act of love, OK?" (Birch Run, Mich., Aug. 11, 2015)

"Jeb Bush has to like the Mexican Illegals because of his wife." (Retweeted and then deleted on Twitter, July 4, 2015)

"I'll win the Latino vote because I'll create jobs. I'll create jobs and the Latinos will have jobs they didn't have." (NBC News, July 8, 2015)

“I’m leading in the Hispanic vote, and I’m going to win the Hispanic vote. I’m also leading in the regular vote.” (Birch Run, Mich., Aug. 11, 2015)

“I don’t have a racist bone in my body.” (“Entertainment Tonight,” July 1, 2015)

“I cherish women. I want to help women. I’m going to be able to do things for women that no other candidate would be able to do … ” (CNN, Aug. 9, 2015)

“I will be so good to women.” (CNN, Aug. 10, 2015)

“Unfortunately, after they are a star, the fun is over for me.” - Donald Trump, who owns the Miss USA pageant, poses on his yacht with 22 state beauty pageant winners in Atlantic City 1988. “I have really given a lot of women great opportunity,” Trump says.

“I will be phenomenal to the women. I mean, I want to help women.” (CBS’ “Face the Nation,” Aug. 9, 2015)

“Oftentimes when I was sleeping with one of the top women in the world I would say to myself, thinking about me as a boy from Queens, ‘Can you believe what I am getting?’” (Think Big: Make it Happen in Business and Life, 2008)

“I’ve never had any trouble in bed …” (Surviving at the Top, 1990)

"I have many women that work for me." (CBS' "Face the Nation," Aug. 9, 2015)

"She's not giving me 100 percent. She's giving me 84 percent, and 16 percent is going towards taking care of children." (TIME, May 23, 2011)

"All of the women on The Apprentice flirted with me—consciously or unconsciously. That's to be expected." (How to Get Rich, 2004)

"I have really given a lot of women great opportunity. Unfortunately, after they are a star, the fun is over for me." (ABC's "Primetime Live," March 10, 1994)

"When a man leaves a woman, especially when it was perceived that he has left for a piece of ass—a good one!—there are 50 percent of the population who will love the woman who was left." (Vanity Fair, September 1990)

"You know who's one of the great beauties of the world, according to everybody? And I helped create her. Ivanka. My daughter, Ivanka. She's 6 feet tall, she's got the best body. She made a lot money as a model—a tremendous amount." (The Howard Stern Show, 2003)

"Every guy in the country wants to go out with my daughter." (New York magazine, Dec. 13, 2004)

"… she does have a very nice figure. I've said if Ivanka weren't my daughter, perhaps I'd be dating her." (ABC's "The View," March 6, 2006)

"I've known Paris Hilton from the time she's 12. Her parents are friends of mine, and, you know, the first time I saw her, she walked into the room and I said, 'Who the hell is that?' … Well, at 12, I wasn't interested. I've never been into that. They're sort of always stuck around that 25 category." (The Howard Stern Show, 2003)

"There's nothing I love more than women, but they're really a lot different than portrayed. They are far worse than men, far more aggressive … " (The Art of the Comeback, 1997)

"If Hillary Clinton can't satisfy her husband what makes her think she can satisfy America?" (Twitter, April 16, 2015)

On Women: "You have to treat 'em like shit." (New York magazine, Nov. 9, 1992)

"What I say is what I say." (Republican presidential debate, Aug. 6, 2015)

"One thing I've learned about the press is that they're always hungry for a good story, and the more sensational the better. … The point is that if you are a little different, or a little outrageous, or if you do things that are bold or controversial, the press is going to write about you." (The Art of the Deal, 1987)

"Sometimes they write positively, and sometimes they write negatively. But from a pure business point of view, the benefits of being written about have far outweighed the drawbacks." (The Art of the Deal, 1987)

"Sometimes it pays to be a little wild." (The Art of the Deal, 1987)

"It's always good to do things nice and complicated so that nobody can figure it out." (New Yorker, May 19, 1997)

"If I get my name in the paper, if people pay attention, that's what matters." (Donald Trump: Master Apprentice, 2005)

"The press portrays me as a wild flamethrower. In actuality, I think I'm much different from that. I think I'm totally inaccurately portrayed." (New Yorker, May 19, 1997)

"You know, it really doesn't matter what they write as long as you've got a young and beautiful piece of ass." (Esquire, 1991)

"There are two publics as far as I'm concerned. The real public and then there's the New York society horseshit. The real public has always liked Donald Trump. The real public feels that Donald Trump is going through Trump-bashing. When I go out now, forget about it. I'm mobbed. It's bedlam." (Vanity Fair, September 1990)

"Controversy, in short, sells." (The Art of the Deal, 1987)

"The show is 'Trump.' And it is sold-out performances everywhere." (Playboy, March 1990)

"I play to people's fantasies. People may not always think big themselves, but they can still get very excited by those who do. That's why a little hyperbole never hurts. People want to believe that something is the biggest and the greatest and the most spectacular." (The Art of the Deal, 1987)

"Most people think small, because most people are afraid of success, afraid of making decisions, afraid of winning. And that gives people like me a great advantage." (The Art of the Deal, 1987)

"You want to know what total recognition is? I'll tell you how you know you've got it. When the Nigerians on the street corners who don't speak a word of English, who have no clue, who're selling watches for some guy in New Jersey—when you walk by and those guys say, 'Trump! Trump!' That's total recognition." (New Yorker, May 19, 1997)

"There is something crazy, hot, a phenomenon out there about me, but I'm not sure I can define it and I'm not sure I want to. How do you think 'The Apprentice' would have done if I wasn't a part of it? There are a lot of imitators now and we'll see how they'll do, but I think they'll crash and burn." (New York Times, Sept. 8, 2004)

"Other rich people don't do commercials because no one asks them. It's just like 'The Apprentice.' I can't tell you how many

of my rich friends are dying, dying to have me put them on that show." (New York Times, Aug. 11, 2004)

"There's something very seductive about being a television star." (TrumpNation: The Art of Being The Donald, 2005)

"I can't help it that I'm a celebrity. What am I going to do, hide under a stone?" (USA Today, Feb. 27, 2004)

"I want a very good-looking guy to play me." (Master Apprentice, 2005)

"Nothing wrong with ego." (Playboy, March 1990)

"Show me someone without an ego, and I'll show you a loser." (Facebook, Dec. 9, 2013)

"One of the key problems today is that politics is such a disgrace. Good people don't go into government. I'd want to change that." (The Advocate, Feb. 15, 2000)

"Our leaders are stupid. They are stupid people. It's just very, very sad." (Las Vegas, April 28, 2011)

"I would hate to think that people blame me for the problems of the world. Yet people come to me and say, 'Why do you allow homelessness in the cities?' as if I control the situation. I am not somebody seeking office." (Playboy, March 1990)

"It's very possible that I could be the first presidential candidate to run and make money on it." (Fortune, April 3, 2000)

"I have no intention of running for President." (TIME, Sept. 14, 1987)

"I don't want to be President. I'm 100 percent sure. I'd change my mind only if I saw this country continue to go down the tubes." (Playboy, March 1990)

"Well, if I ever ran for office, I'd do better as a Democrat than as a Republican—and that's not because I'd be more liberal, because I'm conservative. But the working guy would elect me. He likes me." (Playboy, March 1990)

President Trump "would believe very strongly in extreme military strength. He wouldn't trust anyone. He wouldn't trust the Russians; he wouldn't trust our allies; he'd have a huge military arsenal, perfect it, understand it." (Playboy, March 1990)

"Overseas, we build a school, we build a road, they blow up the school, we build another school, we build another road they blow them up, we build again, in the meantime we can't get a fucking school in Brooklyn." (Las Vegas, April 28, 2011)

"What does it all mean when some wacko over in Syria can end the world with nuclear weapons?" (New York Times, April 8, 1984)

"I think if this country gets any kinder or gentler, it's literally going to cease to exist." (Playboy, March 1990)

"I see the values of this country in the way crime is tolerated, where the people are virtually afraid to say, 'I want the death penalty.' Well, I want it. Where has this country gone when you're not supposed to put in a grave the son of a bitch who robbed, beat, murdered and threw a 90-year-old woman off the building?" (Playboy, March 1990)

"I know politicians who love women who don't even want to be known for that, because they might lose the gay vote, OK?" (Playboy, March 1990)

"Politicians are all talk and no action." (Twitter, May 27, 2015)

"Politicians are all talk and no action." When Donald Trump enters the ring to shave the head of World Wrestling Entertainment chairman Vince McMahon in 2007.

"When you need zone changes, you're political. ... You know, I'll support the Democrats, the Republicans, whatever the hell I have to support." (BuzzFeed, Feb. 13, 2014)

"I will tell you that our system is broken. I gave to many people. Before this, before two months ago, I was a businessman. I give to everybody. When they call, I give. And do you know what? When I need something from them two

years later, three years later, I call them—they are there for me." (Republican presidential debate, Aug. 6, 2015)

"Power corrupts." (Playboy, March 1990)

"Part of the beauty of me is that I am very rich." (ABC's "Good Morning America," March 17, 2011)

"I'm really rich." (New York City, June 16, 2015)

"Look at these contracts. I get these to sign every day. I've signed hundreds of these. Here's a contract for $2.2 million. It's a building that isn't even opened yet. It's eighty-three percent sold, and nobody even knows it's there. For each contract, I need to sign twenty-two times, and if you think that's easy ... You know, all the buyers want my signature. I had someone else who works for me signing, and at the closings the buyers got angry. I told myself, 'You know, these people are paying a million eight, a million seven, two million nine, four million one—for those kinds of numbers, I'll sign the fucking contract.' I understand. Fuck it. It's just more work." (New Yorker, May 19, 1997)

"My attitude is if somebody's willing to pay me $225,000 to make a speech, it seems stupid not to show up. You know why I'll do it? Because I don't think anyone's ever been paid that much." (New Yorker, May 19, 1997)

"I look very much forward to showing my financials, because they are huge." (TIME, April 14, 2011)

"I have a total net worth and now with the increase it will be well over $10 billion, but here total net worth of $8 billion. Net worth—not assets, not liabilities—a net worth. ... I'm not doing that to brag. Because you know what? I don't have to brag. I don't have to. Believe it or not." (New York City, June 16, 2015)

"People say the '80s are dead, all the luxury, the extravagance. I say, 'What?' Am I supposed to change my taste because it's a new decade? That's bullshit." (Playboy, May 1997)

"If you don't tell people about your success, they probably won't know about it." (How to Get Rich, 2004)

"I know how to sell. Selling is life. You can have the greatest singer in the world, but if nobody knows who he is, he'll never have the opportunity to sing." (Sports Illustrated, Feb. 13, 1984)

"There are singers in the world with voices as good as Frank Sinatra's, but they're singing in their garages because no one has ever heard of them. You need to generate interest, and you need to create excitement." (The Art of the Deal, 1987)

"I've always felt that a lot of modern art is a con, and that the most successful painters are often better salesmen and promoters than they are artists." (The Art of the Deal, 1987)

"That's why the banks love me. They love my reputation." (New York Times, March 28, 2004)

"I really value my reputation and I don't hesitate to sue." (Village Voice, Jan. 15, 1979)

"I think the brand is huge. What is it about me that gets Larry King his highest ratings?" (TrumpNation: The Art of Being The Donald, 2005)

"My brand became more famous as I became more famous, and more opportunities presented themselves." (Amazon.com, 2007)

"Because I've been successful, make money, get headlines, and have authored bestselling books, I have a better chance to make my ideas public than do people who are less well known." (The America We Deserve, 2000)

"Let's say I was worth $10. People would say, 'Who the [expletive] are you?' You understand? They know my statement. Fortune. My book, The Art of the Deal, based on my fortune. If I didn't make a fortune, who the [expletive] is going to buy The Art of the Deal? That's why they watched 'The Apprentice.' Because of my great success." (Washington Post, July 12, 2015)

"People think I'm a gambler. I've never gambled in my life. To me, a gambler is someone who plays slot machines. I prefer to own slot machines. It's a very good business being the house." (The Art of the Deal, 1987)

"I like the casino business. I like the scale, which is huge, I like the glamour, and most of all, I like the cash flow." (The Art of the Deal, 1987)

"Cash is king, and that's one of the beauties of the casino business." (Playboy, March 1990)

"How much have I made off the casinos? Off the record, a lot." (Master Apprentice, 2005)

"And while I can't honestly say I need an 80-foot living room, I do get a kick out of having one." (The Art of the Deal, 1987)

"Let me tell you something about the rich. They have a very low threshold for pain." (New York magazine, Feb. 11, 1985)

"Fighting for the last penny is a very good philosophy to have." (Esquire, January 2004)

"I aim very high, and then I just keep pushing and pushing and pushing to get what I'm after." (The Art of the Deal, 1987)

"I don't like to lose." (New York Times, Aug. 7, 1983)

"I'll do nearly anything within legal bounds to win." (The Art of the Deal, 1987)

"If you don't win you can't get away with it. And I win, I win, I always win. In the end, I always win, whether it's in golf, whether it's in tennis, whether it's in life, I just always win. And I tell people I always win, because I do." (TrumpNation: The Art of Being The Donald, 2005)

"I'll do what I have to do." (Atlantic, April 2013)

"I do whine because I want to win, and I'm not happy about not winning, and I am a whiner, and I keep whining and whining until I win." (CNN, Aug. 10, 2015)

"Hey, look, I had a cold spell from 1990 to '91. I was beat up in business and in my personal life. But you learn that you're either the toughest, meanest piece of shit in the world or you just crawl into a corner, put your finger in your mouth, and say, 'I want to go home.' You never know until you're under pressure how you're gonna react. Guys that I thought were tough were nothin'." (New York magazine, Aug. 15, 1994)

"The mind can overcome any obstacle. I never think of the negative." (New York Times, Aug. 7, 1983)

"It's been said that I believe in the power of positive thinking. In fact, I believe in the power of negative thinking." (The Art of the Deal, 1987)

"A lot of people sit down and discuss their lives, things like are they happy, but it's not like that with me. I don't think positively, I don't think negatively, I just think about the goal. But it's not like I sit down and write goals. I just do things." (Master Apprentice, 2005)

"If you asked Babe Ruth how he hit home runs, he was unable to tell you. I do things by instinct." (New York Times, Sept. 8, 2004)

"A lot of people build a brand and they study it very carefully and every move is calculated. My moves are not calculated. My moves are totally uncalculated." (TrumpNation: The Art of Being The Donald, 2005)

"For many years I've said that if someone screws you, screw them back. When somebody hurts you, just go after them as viciously and as violently as you can." (How to Get Rich, 2004)

"The way I see it, critics get to say what they want about my work, so why shouldn't I be able to say what I want to about theirs?" (The Art of the Deal, 1987)

"Sometimes, part of making a deal is denigrating your competition." (The Art of the Deal, 1987)

"One of the problems when you become successful is that jealousy and envy inevitably follow. There are people – I categorize them as life's losers—who get their sense of accomplishment and achievement from trying to stop others. As far as I'm concerned, if they had any real ability they wouldn't be fighting me, they'd be doing something constructive themselves." (The Art of the Deal, 1987)

“Rosie O’Donnell’s disgusting. I mean, both inside and out. You take a look at her, she’s a slob … ” (“Entertainment Tonight,” Dec. 21, 2006)

Among Trump's statements on various public women: “Rosie O'Donnell's disgusting”; Bette Midler is “grotesque”; Arianna Huffington is “a dog"; Cher, however, is only “somewhat of a loser.” | AP Photos.

“She called me a snake oil salesman, and, you know, coming from Rosie, that’s pretty low, because when you look at her, and when you see the mind, the mind is—is weak. I don’t see it. I don’t get it. I never understood. How does she even get on television?” (“Entertainment Tonight,” Dec. 21, 2006)

“Rosie’s a person that’s very lucky to have her girlfriend. And she better be careful or I’ll send one of my friends over to pick up her girlfriend. Why would she stay with Rosie if she had another choice?” (“Entertainment Tonight,” Dec. 21, 2006)

“Probably I’ll sue her. Because it would be fun. I’d like to take some money out of her fat-ass pockets.” (“Entertainment Tonight,” Dec. 21, 2006)

“I think she’s a terrible person. I can look at people and see what they are.” (New York Post, December 22, 2006)

Bette Midler is “grotesque.” (Twitter, Oct. 28, 2012)

Arianna Huffington is “a dog.” (Twitter, April 6, 2015)

"Angelina Jolie is sort of amazing because everyone thinks she's like this great beauty. And I'm not saying she's an unattractive woman, but she's not a beauty, by any stretch of the imagination. I really understand beauty." (CNN, Oct. 9, 2006)

"Karl Rove is a total loser." (Twitter, Feb. 7, 2013)

Republican pollster Frank Luntz is a "total loser!" (Twitter, Aug. 3, 2014)

Russell Brand is a "major loser!" (Twitter, Oct. 16, 2014)

"Cher is somewhat of a loser. She's lonely. She's unhappy. She's very miserable." (Fox News, May 15, 2012)

George Will is a "moron." (Twitter, April 17, 2015)

Chuck Todd is a "moron." (Twitter, Aug. 9, 2013)

Former Secretary of Defense Chuck Hagel: "Obvious moron." (Twitter, Aug. 23, 2014)

Megyn Kelly is a "bimbo." (Twitter, Aug. 7, 2015)

Michelle Malkin is a "dummy." (Twitter, Oct. 25, 2012)

Brian Williams is a "dummy." (Twitter, March 6, 2013)

Graydon Carter? "Dummy." (Twitter, Dec. 19, 2012)

John McCain is "not a war hero. ... He is a war hero because he was captured. I like people who weren't captured, OK?" (Ames, Iowa, July 18, 2015)

"Truly weird Senator Rand Paul of Kentucky reminds me of a spoiled brat without a properly functioning brain." (Twitter, Aug. 10, 2015)

"I just realized that if you listen to Carly Fiorina for more than ten minutes straight, you develop a massive headache. She has zero chance!" (Twitter, Aug. 9, 2015)

"What a stiff, what a stiff, Lindsey Graham." (Bluffton, S.C., July 21, 2015)

Rick Perry "put on glasses so people think he's smart. ... People can see through the glasses." (Bluffton, S.C., July 21, 2015)

Rick Santorum? "I have a big plane. He doesn't." (Des Moines Register, April 8, 2015)

The record ratings for the Republican debate earlier this month? "Who do you think they're watching? Jeb Bush? Huh? I don't think so." (Birch Run, Mich., Aug. 11, 2015)

And George W. Bush is "no Einstein." (Minneapolis, Jan. 7, 2000)

Bernie Sanders and Black Lives Matter protesters in Seattle? "I would never give up my microphone! I thought that was disgusting. That showed such weakness, the way he was taken away by two young women. The microphone!" (Birch Run, Mich., Aug. 11, 2015)

"My favorite part (of Pulp Fiction) is when Sam has his gun out in the diner and he tells the guy to tell his girlfriend to shut up. Tell that bitch to be cool. Say: Bitch be cool. I love those lines." (TrumpNation: The Art of Being The Donald, 2005)

"I'm not the world's happiest person." (New York Magazine, March 5, 1990)

"I think I am a nice person. People who know me like me." (New York City, June 16, 2015)

"I can be a killer and a nice guy. You have to be everything. You have to be strong. You have to be sweet. You have to be ruthless. And I don't think any of it can be learned. Either you have it or you don't. And that is why most kids can get straight A's in school but fail in life." (Playboy, March 1990)

"I like the challenge and tell the story of the coal miner's son. The coal miner gets black-lung disease, his son gets it, then his son. If I had been the son of a coal miner, I would have left

the damn mines. But most people don't have the imagination—or whatever—to leave their mine. They don't have 'it.'" (Playboy, March 1990)

"It is an ability to become an entrepreneur, a great athlete, a great writer. You're either born with it or you're not." (Playboy, March 1990)

"But when somebody tries to sucker punch me, when they're after my ass, I push back a hell of a lot harder than I was pushed in the first place. If somebody tries to push me around, he's going to pay a price. Those people don't come back for seconds. I don't like being pushed around or taken advantage of. And that's one of the problems with our country today. This country is being pushed around by everyone." (Playboy, March 1990)

"Sorry losers and haters, but my I.Q. is one of the highest — and you all know it!" (Twitter, May 8, 2013)

"... of course, it's very hard for them to attack me on looks, because I'm so good looking." (NBC's "Meet the Press," Aug. 9, 2015)

"I can never apologize for the truth." (Fox News, July 5, 2015)

"Who knows what's in the deepest part of my mind?" (BuzzFeed, Feb. 13, 2014)

"I enjoy testing friendships." (Playboy, March 1990)

"I do love provoking people." (BuzzFeed, Feb. 13, 2014)

"You can never tell until you test; the human species is interesting in that way." (Playboy, March 1990)

"Everything in life to me is a psychological game, a series of challenges you either meet or don't." (Playboy, March 1990)

"My father was a builder in Brooklyn and Queens, a very smart businessperson who understood life." (Esquire, January 2004)

"We had a relationship that was almost businesslike." (The Art of the Deal, 1987)

"Not many sons have been able to escape their fathers." (New York Times, Aug. 7, 1983)

"Look, I had friends whose fathers were very successful, and the fathers were jealous of the sons' success and tried to hurt them, keep them down, because they wanted to be the king. My father was the exact opposite. He used to carry around articles (about me)." (New York Times, Jan. 2, 2000)

"I was always very much accepted by my father. He adored Donald Trump ... " (Playboy, March 1990)

"He was a strong, strict father, a no-nonsense kind of guy, but he didn't hit me." (Playboy, March 1990)

"My father taught me to keep my guard up. The world is a pretty vicious place." (Esquire, January 2004)

"People are too trusting. I'm a very untrusting guy." (Playboy, March 1990)

"I believe that, unfortunately, people are out for themselves." (Playboy, March 1990)

"Statistically, my children have a very bad shot. Children of successful people are generally very, very troubled, not successful. They don't have the right shtick. You never know until they're tested." (Playboy, March 1990)

"I want five children, like in my own family, because with five, then I will know that one will be guaranteed to turn out like me." (Vanity Fair, September 1990)

" … when you're rich, you can have as many kids as you want. Being rich makes it easier to have kids." (TrumpNation: The Art of Being The Donald, 2005)

"The hardest thing for me about raising kids has been finding the time. I know friends who leave their business so they can spend more time with their children, and I say, 'Gimme a break!'" (New York magazine, Dec. 13, 2004)

"My marriage, it seemed, was the only area of my life in which I was willing to accept something less than perfection." (Surviving at the Top, 1990)

"I would never buy Ivana any decent jewels or pictures. Why give her negotiable assets?" (Vanity Fair, September 1990)

And Marla, wife No. 2? "I was bored when she was walking down the aisle. I kept thinking: What the hell am I doing here? I was so deep into my business stuff. I couldn't think of anything else." (TrumpNation: The Art of Being The Donald, 2005)

"It's all in the hunt and once you get it, it loses some of its energy. I think competitive, successful men feel that way about women. Don't you agree? Really, don't you agree?" (TrumpNation: The Art of Being The Donald, 2005)

"Marla was always wanting me to spend more time with her. 'Why can't you be home at 5 o'clock like other husbands?' she would ask. Sometimes, when I was in the wrong mood, I would give a very materialistic answer. 'Look, I like working. You don't mind traveling around in beautiful helicopters and airplanes, and you don't mind living at the top of Trump Tower, or at Mar-a-Lago, or traveling to the best hotels, or shopping in the best stores and never having to worry about money, do you?'" (The Art of the Comeback, 1997)

" ... I'm married to my business. It's been a marriage of love. So, for a woman, frankly, it's not easy in terms of relationships. But there are a lot of assets." (New York magazine, Dec. 13, 2004)

"But, I think you understand, I don't have very much time. I just don't have very much time. There's nothing I can do about what I do other than stopping. And I just don't want to stop." (New Yorker, May 19, 1997)

"... you need love, you need trust, you need sex, you need lots of different things—all of which are very complex." (Esquire, January 2004)

"For me, business comes easier than relationships." (Esquire, January 2004)

Trump and father Fred, at right, in an undated photograph. Says Donald Trump of his father: "He adored Donald Trump." | Photo by Dennis Caruso/Getty Images

"I don't do it for the money." (The Art of the Deal, 1987)

"I do it to do it." (The Art of the Deal, 1987)

"I've done an incredible job." (Atlantic, April 2013)

"I nod, and it is done." (Esquire, January 2004)

"Know what? After shaking five thousand hands, I think I'll go wash mine." (New Yorker, May 19, 1997)

"I don't dwell on the past." (Playboy, March 1990)

"I don't look forward or not look forward." (Washington Post, July 12, 2015)

"I always go into the center." (New Yorker, May 19, 1997)

"You gotta say, I cover the gamut. Does the kid cover the gamut? Boy, it never ends. I mean, people have no idea. Cool life. You know, it's sort of a cool life." (New Yorker, May 19, 1997)

"I've had a lot of victories. I fight hard for victory, and I think I enjoy it as much as I ever did. But I realize that maybe new victories won't be the same as the first couple." (People, July 9, 1990)

"... the same assets that excite me in the chase, often, once they are acquired, leave me bored. For me, you see, the important thing is the getting, not the having." (Surviving at the Top, 1990)

"I truly believe that someone successful is never really happy, because dissatisfaction is what drives him." (Playboy, March 1990)

"I rarely stop for lunch." (The Art of the Deal, 1987)

"I don't sleep more than four hours a night." (Playboy, March 1990)

“I’m a guy who lies awake at night and thinks and plots.” (New York magazine, Nov. 9, 1992)

“Some people cast shadows, and other people choose to live in those shadows.” (New York Times, Sept. 11, 2005)

“Did you know that New York Construction News named Donald Trump the developer and owner of the year?” (Fortune, April 3, 2000)

“How long is your article?” (Vanity Fair, September 1990)

“Is it a cover?” (Vanity Fair, September 1990)

“What do people say about me? Do they say I’m loyal? Do they say I work hard?” (Village Voice, Jan. 15, 1979)

"Make America Great Again!"

2016 Presidential Campaign Slogan

“I will be the greatest jobs president that God ever created.”

Campaign rally, 2016

"I have a great relationship with the blacks. I’ve always had a great relationship with the blacks."

***The New York Times* interview, 2011**

"Sorry, there is no STAR on the stage tonight! So don't worry, they are 'somewhat' harmless, but you need to relax!"

Trump Tweet, 2016

"I could stand in the middle of Fifth Avenue and shoot somebody, and I wouldn't lose any voters."

Campaign rally in Sioux Center, Iowa, 2016

"When somebody is the president of the United States, the authority is total, and that's the way it's got to be."

***Time* Magazine interview, 2019**

"The United States is going to be so great again that you won't believe it."

Campaign rally, 2016

"I'm not a politician. I'm a businessman."

***Fox News* interview, 2016**

"You know, it's not about me, it's about us."

2016 Presidential Campaign Speech

"It's going to be only America first, America first."

Inaugural Address, January 20, 2017

"Nobody has more respect for women than I do."

***The View*, 2015**

"I have no problem with being a showman. I think it's part of who I am."

Interview with *The New York Times*, 2016

"We're going to build a wall, and Mexico is going to pay for it."

Campaign rally, 2015

"I'm the least racist person you'll ever meet."

***The New York Times* interview, 2019**

"I know more about ISIS than the generals do. Believe me."

Campaign rally, 2016

"Nobody knew healthcare could be so complicated."

***Time* Magazine, 2017**

"I don't want to be a president that's just a president of the people who voted for me. I want to be a president of all the people."

Inaugural Address, January 20, 2017

"I will build a great wall — and nobody builds walls better than me, believe me."

***Fox News* interview, 2015**

"I think I'm a very stable genius."

Tweet, January 6, 2018

"I'm a very stable genius."

Washington Post interview, 2018

"I'm not a loser. I've had great success. I've built a great company."

The New York Times interview, 2011

"I've said if Ivanka weren't my daughter, perhaps I'd be dating her."

The Howard Stern Show, 2003

"Take a look. You take a look. Look at her. Look at her words. Tell me what you think. I don't think so." regarding journalist Natasha Stoynoff attractiveness after she said Trump had pushed her against a wall and forcibly tried to kiss her during an interview in 2005.

"I will be so good at the military your head will spin."

Campaign rally, 2016

"I've always been very good at doing deals. That's why I'm president."

Fox News interview, 2017

"The beauty of me is that I'm very rich."

The New York Times, 2007

"Sometimes by losing a battle, you find a new way to win the war."

The Art of the Deal, 1987

"I know words. I have the best words."

Campaign rally, 2015

"I don't care about the press. I don't care about what they say. I know what I'm doing."

Interview with *The New York Times*, 2017

"I don't think about losing. I only think about winning."

Campaign rally, 2016

"I am a big fan of the police. I think the police do a great job."

Fox News interview, 2016

"The only thing I know for sure is that I'm right, and everybody else is wrong."

The Art of the Deal, 1987

"We will make America safe again. We will make America great again."

Republican National Convention, 2016

"The press is the enemy of the people."

Tweet, February 17, 2017

"It's very hard to make a deal when you're always looking over your shoulder."

Interview with *Fox News*, 2017

"I'll be back. I'll be back and I'll be stronger than ever."

Rally speech, 2021

"They're not after me. They're after you. I'm just in the way."

Campaign rally, 2019

"I will never let you down."

2016 Presidential Campaign Speech

"There is no greater feeling than being a winner."

The Art of the Deal, 1987

"You know, I think you have to be able to laugh at yourself, especially when you're in a public position."

The New York Times, 2017

"I don't think the government should be doing everything. I don't want to live in a country where the government is always telling me what to do."

Fox News interview, 2017

"I've had a lot of people tell me that I have a great memory. They're amazed by it."

Campaign rally, 2016

"You know, I think the world of China. They're one of the great countries, and they're doing an amazing job."

CNN interview, 2017

"I'm not a politician. I'm a businessman. That's the difference."

NBC News interview, 2016

"You can't be a good president unless you're a good negotiator."

Fox News interview, 2017

"I like people who weren't captured."

Campaign rally, 2015 (referring to John McCain, war hero and POW survivor)

"I am the best thing that's ever happened to the Secret Service."

The New York Times, 2017

"The future belongs to the dreamers, not to the cynics."

2016 Campaign rally

"I don't need anybody's approval."

Fox News interview, 2016

"It's very important to treat people with respect, but it's also important to know when to fight back."

The Art of the Deal, 1987

"I'm a great negotiator. I've done it all my life."

Campaign rally, 2016

"We are going to win so much, you're going to be so sick and tired of winning."

Campaign rally, 2016

"The only thing I'm interested in is success, and if I can do something that's going to help this country, I'm all for it."

The New York Times, 2016

"I'm the most transparent president ever."

The New York Times interview, 2017

"I'm a businessman, I'm not a politician, I know how to negotiate, I know how to make deals."

Campaign rally, 2015

"I don't like to say, 'I told you so,' but I told you so."

The New York Times, 2016

"I'm very proud of my hair. I've had it for a long time, and I've had people ask me, 'How do you do that?'"

***The Howard Stern Show*, 2003**

"I never want to be a politician. That's the worst thing in the world."

Interview with *NBC News*, 2015

"I think I know more about ISIS than the generals do, believe me."

***NBC News* interview, 2016**

"I will be the greatest president God ever created."

***The New York Times* interview, 2011**

"The problem with the Republicans is that they don't fight hard enough. They're not tough enough."

Campaign rally, 2016

"I could shoot somebody and I wouldn't lose voters."

Campaign rally in Sioux Center, Iowa, 2016

"I'm very proud of what we've done. We've accomplished a lot."

***The Washington Post*, 2017**

"The Democrats are the party of crime."

Campaign rally, 2016

"I'm not saying I have the best judgment, but I have the best people."

The Washington Post, 2017

"I will bring jobs back to this country. Jobs are the most important thing."

Campaign rally, 2016

"You have to show people who's boss."

The Art of the Deal, 1987

"If I had to, I could be a great politician. I could be a great president, but I'm not doing that. I'm doing it for you."

Campaign rally, 2016

"I want to make America great again. It's as simple as that."

2016 Presidential Campaign Speech

"I have a lot of respect for the people who work in this country, but I also believe in doing things my way."

NBC News interview, 2017

"I think I'm doing a great job, and I think the American people agree with me."

The Washington Post interview, 2017

"I've made billions and billions of dollars. I've employed thousands and thousands of people."

The Art of the Deal, 1987

"I love people who are great at negotiating, and I think I'm the best at it."

Fox News interview, 2016

"We're going to make America great again. Believe me, it's going to happen."

2016 Presidential Campaign Speech

"I've done a lot of things that people thought were impossible, and I'm going to do a lot more."

The New York Times, 2017

"I'm not a traditional politician. I think that's a good thing."

Fox News interview, 2016

"I think the thing with Washington is they don't like change. And I am a big fan of change."

NBC News interview, 2017

"I am the king of the deal. No one can make deals better than I can."

Campaign rally, 2016

"They say it's impossible to do this, but I'm going to do it."

The Washington Post, 2017

"My whole life has been a series of deals. Everything's a deal, I do deals."

The Art of the Deal, 1987

"I'm the only one who can fix this country."

2016 Presidential Campaign Speech

"It's an honor to be the president of the United States."

Washington Post, 2017

"I know more about drones than anybody. I know about everything."

NBC News interview, 2016

"I'm the best thing that's ever happened to the military."

The Washington Post, 2017

"Our country is doing well. We're having a lot of success."

Fox News interview, 2017

"I don't make mistakes. I never make mistakes."

The New York Times, 2016

"I've done more in two years than most people have done in a lifetime."

Campaign rally, 2018

“We have to be tough on crime. We have to be tough on criminals.”

***Fox News* interview, 2016**

“You need a leader who can make decisions quickly. I’m that leader.”

Source: Campaign rally, 2016

"The wall is going to be built. The wall is going to happen."

Source: *Fox News* interview, 2017

"I’ve been great with women. Everybody says that."

Source: *Fox News* interview, 2016

“I’m not going to tell you how I’m going to negotiate with China. I’ll let you know after I do it.”

***The New York Times*, 2016**

“People are saying that I am a great president, but I’m very humble.”

***Fox News* interview, 2018**

“I’ll tell you what, you get me the right people, and we’ll have a great economy.”

Campaign rally, 2016

“I have the best people. I hire only the best people.”

***The Washington Post*, 2016**

"I'm a person that wants to bring people together."

***The New York Times* interview, 2016**

“I’m going to be the best job president that God ever created.”

Campaign rally, 2016

“I know how to get things done. I know how to make deals.”

***NBC News* interview, 2016**

"We’ve made so much progress in the last two years. It’s incredible.”

***The Washington Post* interview, 2018**

“I love fighting the fake news.”

***Fox News* interview, 2017**

"I think the word ‘fake news’ is a great term. I think it’s a great term because it’s an honest term."

***Fox News* interview, 2017**

"I’ve had so many people tell me, ‘You know, Donald, you could have made a lot more money, but you’re too busy doing other things. You’re doing things for this country.’”

The Washington Post interview, 2017

"I've done more for this country than anybody else has done in recent history."

Fox News interview, 2017

"I think I'm a very good judge of people. I've been doing it my whole life."

The Washington Post, 2016

"The fake news media is out of control. They lie, they exaggerate, they distort."

Fox News interview, 2017

"I know the best people. I know the best deals. Believe me."

NBC News interview, 2016

"If you're looking for someone who's politically correct, you've got the wrong guy."

Campaign rally, 2015

"If you don't want me to talk about the media, you shouldn't let them treat me the way they do."

NBC News interview, 2017

"I don't like to be called a politician because I'm not a politician."

NBC News interview, 2017

“You’re going to see things that are going to surprise people. It’s going to be beautiful.”

Campaign rally, 2016

“I’m a very optimistic person. I think our future is bright.”

Fox News interview, 2017

“I want to make America great again. You know that.”

Campaign rally, 2016

“I don’t think I’ve ever seen a president who works harder than I do.”

Fox News interview, 2017

“I’ve always had a great relationship with people from all walks of life.”

The Washington Post, 2016

"I want to keep this country safe. That’s my job."

NBC News interview, 2017

"I think I’ve done a great job with the military."

The New York Times, 2017

"We’ve had the best economy. You wouldn’t believe how great it is."

Fox News interview, 2017

"I'm very proud of my company. We've built incredible buildings."

The Art of the Deal, 1987

"I think there's a reason I'm winning. It's because I've been consistent."

Campaign rally, 2016

"I am a very loyal person, and I expect loyalty in return."

The Washington Post, 2016

"We will not let people die on the streets. We will not let people be taken advantage of anymore."

Fox News interview, 2017

"We've made so much progress in the last two years. It's been tremendous."

The New York Times, 2018

"You don't have to like me, but I am making this country great again."

Campaign rally, 2016

"I'm a big fan of the military. I think they do a great job."

The Washington Post, 2017

"I'll do what I have to do to protect this country, and I don't care what anyone thinks."

The New York Times, 2016

"People are really starting to like me. They really are."

Fox News interview, 2017

"There's a reason people come to this country. It's because it's the best country in the world."

NBC News interview, 2017

"I love it when people say that I'm not qualified. That's the best thing."

The New York Times, 2016

"I'll put America first every time, and that's why I was elected."

Fox News interview, 2017

"If you believe in what's right, if you believe in yourself, you'll succeed."

The Washington Post, 2017

"I'm not afraid of anybody."

Campaign rally, 2016

"I'll be the first to admit that I don't know everything, but I know how to get the best people around me."

The New York Times, 2016

"I know how to win. I've been doing it for years."

The Art of the Deal, 1987

"I don't care what they say. I know what I'm doing."

NBC News interview, 2017

"We're going to have the best healthcare system. Believe me."

Fox News interview, 2017

"I'm going to keep doing what's right for this country, no matter who likes it or not."

The Washington Post, 2017

"I'm a winner. I'm always a winner."

Campaign rally, 2016

"Lock her up."

Multiple campaign rallies on imprisoning Hillary Clinton his opponent for presidential election, 2016

Trump said he had decided that imprisoning Clinton "would have been a terrible thing." Then he added: "I didn't say 'lock her up,' but the people would all say 'lock her up, lock her up'."

Fox News, June 2024

"I'll never let you down."

Campaign rally, 2016

Trump called Cheney "a deranged person" and added, "But the reason she couldn't stand me is that she always wanted to go to war with people. If it were up to her we'd be in 50 different countries. She's a radical war hawk. Let's put her with the rifle standing there with nine barrels shooting at her. OK, let's see how she feels about it. You know, when the guns are trained on her face."

Fox News, 2024

"They're eating dogs."

Lying about Haitians during a Presidential Debate with Kamala Harris, September 2024

"I'm not here to make friends. I'm here to get the job done."

The Washington Post, 2016

"You watch, we're going to win so much that you'll be begging us to stop winning."

Campaign rally, 2016

"Elon [Musk]knows those computers better than anybody. All those computers. Those vote-counting computers. And we ended up winning Pennsylvania like in a landslide."

Tweet, January 2025

"Christians, get out and vote, just this time. "You won't have to do it anymore. Four more years, you know what, it will be fixed, it will be fine, you won't have to vote anymore, my beautiful Christians . . . I love you Christians. I'm a Christian. I love you, get out, you gotta get out and vote. In four years, you don't have to vote again, we'll have it fixed so good you're not going to have to vote."

Turning Point Action in West Palm Beach, Florida, 2024

"In many cases, I didn't respect them," Trump said when asked why he decided to ignore his lawyers and advisers who told him he lost the 2020 election to Joe Biden. "But I did respect others. I respected many others that said the election was rigged."

Meet the Press, September 2023

"I don't think you're going to have another election in this country, if we don't win this election … certainly not an election that's meaningful."

Campaign rally, Ohio, March 2024

"Well, he's unleashed something that everybody, we've all known about this for a hundred years," [in reference to President Biden and his administration]. We've watched other countries do it and, in some cases, effective and in other cases, the country's overthrown or it's been totally ineffective. But we've watched this for a long time, and it's not unique, but it's unique for the United States. Yeah. If they do this and they've already done it, but if they want to follow through on this, yeah, it could certainly happen in reverse. It could certainly happen in reverse. What they've done is they've released the genie out of the box . . . "They call it weaponization [of the Department of Justice against political enemies], and the people aren't going to stand for it," Trump said. "But yeah. they have done something that allows the next party. I mean, if somebody — if I happen to be president and I see somebody who's doing well and beating me very badly, I say, 'Go down and indict them.' Mostly what that would be, you know, they would be out of business. They'd be out, they'd be out of the election."

Univision Interview, 2024

“I better use some Tic Tacs just in case I start kissing her. You know I'm automatically attracted to beautiful—I just start kissing them. It's like a magnet. Just kiss. I don't even wait. And when you're a star, they let you do it. You can do anything. Grab 'em by the pussy. You can do anything.”

Recording off-air with Access Hollywood (and Billy Bush)

“I could shoot somebody, and I wouldn’t not lose voters.”

Campaign rally, Iowa, 2016

“No, no, no, other than day one. We’re closing the border, and we’re drilling, drilling, drilling. After that, I’m not a dictator.’”